RESHAPING THE BUSINESS WORLD POST-COVID-19

Management Strategies for Sustainable Behavior Change

21st Century Business Management

RESHAPING THE BUSINESS WORLD POST-COVID-19

Management Strategies for Sustainable Behavior Change

Edited by

Arvind K. Birdie, PhD

Ruchi Joshi, PhD

AAP | APPLE ACADEMIC PRESS

First edition published 2024

Apple Academic Press Inc.
1265 Goldenrod Circle, NE,
Palm Bay, FL 32905 USA

760 Laurentian Drive, Unit 19,
Burlington, ON L7N 0A4, CANADA

CRC Press
2385 NW Executive Center Drive,
Suite 320, Boca Raton FL 33431

4 Park Square, Milton Park,
Abingdon, Oxon, OX14 4RN UK

Apple Academic Press exclusively co-publishes with CRC Press, an imprint of Taylor & Francis Group, LLC

Library and Archives Canada Cataloguing in Publication

Title: Reshaping the business world post-COVID-19 : management strategies for sustainable behavior change / edited by Arvind K. Birdie, PhD, Ruchi Joshi, PhD.
Names: Birdie, Arvind K., editor. | Joshi, Ruchi (Lecturer in psychology), editor.
Description: First edition. | Series statement: 21st century business management book series | Includes bibliographical references and index.
Identifiers: Canadiana (print) 20230445683 | Canadiana (ebook) 20230445802 | ISBN 9781774913406 (hardcover) | ISBN 9781774913413 (softcover) | ISBN 9781003372424 (ebook)
Subjects: LCSH: Organizational behavior. | LCSH: Personnel management. | LCSH: COVID-19 Pandemic, 2020.
Classification: LCC HD58.7 .R47 2024 | DDC 658.3—dc23

Library of Congress Cataloging-in-Publication Data

Names: Birdie, Arvind K., editor. | Joshi, Ruchi, editor.

Title: Reshaping the business world post-COVID-19 : management strategies for sustainable behavior change / edited by Arvind K. Birdie, PhD, Ruchi Joshi, PhD.

Description: Palm Bay, FL : Apple Academic Press, 2024. | Series: 21st century business management | Includes bibliographical references and index. | Summary: "The COVID-19 pandemic dramatically changed our normal-both in life and in business. The timely volume provides a map of how the world has been significantly changed by the post-COVID-19 pandemic, not only in terms of work and business life but also linking other areas of personal life as well. The volume explores the diverse impacts of the pandemic on businesses and workplaces, addressing topics such as changes in organizational structures, operations, and marketing and consumer behavior. Sharing their rich insights and perspectives on today's business world, the authors also look at personal psychological well-being, the role of spirituality, employee satisfaction, an organization's future competitiveness, and quality of life have been affected and changed by the COVID-19 pandemic. Chapters discuss the challenges and complications of flexible and hybrid working styles and digital collaboration platforms such as Microsoft teams. Other topics include changed consumer choices and shopping psychology, internet addiction, mental health challenges, new psychological aspects of the art and culture industry, and more. The authors also share effective strategies for creating work-life balance and improving psychological well-being and for navigating the "new normal." Reshaping the Business World Post-COVID-19: Management Strategies for Sustainable Behavior Change will be valuable for both industry and academia as it covers concepts of business from various perspectives. The book is sure to help managers of all types navigate the new normal"-- Provided by publisher.

Identifiers: LCCN 2023021318 (print) | LCCN 2023021319 (ebook) | ISBN 9781774913406 (hardback) | ISBN 9781774913413 (paperback) | ISBN 9781003372424 (ebook)

Subjects: LCSH: Management. | Organizational change. | Strategic planning. | COVID-19 Pandemic, 2020---Economic aspects.

Classification: LCC HD31.2 .R475 2024 (print) | LCC HD31.2 (ebook) | DDC 658--dc23/eng/20230512

LC record available at https://lccn.loc.gov/2023021318

LC ebook record available at https://lccn.loc.gov/2023021319

ISBN: 978-1-77491-340-6 (hbk)
ISBN: 978-1-77491-341-3 (pbk)
ISBN: 978-1-00337-242-4 (ebk)

About the 21st Century Business Management Book Series

Series Editor:
Arvind K. Birdie, PhD
Associate Professor, School of Humanities, Social Sciences and Education, G. D Goenka University, Sohna, Gurgaon, India

CURRENT BOOKS IN THE SERIES

Employees and Employers in Service Organizations: Emerging Challenges and Opportunities
Editor: Arvind K. Birdie, PhD

The Future of Organizations: Workplace Issues and Practices
Editor: Arvind K. Birdie, PhD

Cross-Cultural Exposure and Connections: Intercultural Learning for Global Citizenship
Editor: Arvind K. Birdie, PhD

Artificial Intelligence Techniques in Human Resource Management
Editors: Soumi Ghosh, PhD, Soumi Majumder, and Santosh Kumar Das, PhD

Reshaping the Business World Post-COVID-19
Editors: Arvind K. Birdie, PhD, and Ruchi Joshi, PhD

Responsible Management Practice for Sustainability
Editors: S. Vasantha, PhD, A. Menaga, Renuka Pushpanjalee Herath, PhD, and Nithya Ramachandran, PhD

Pharma Marketing and Pharmacoeconomics: Opportunities, Challenges, and the Way Forward
Editors: Rishabha Malviya, PhD, Pramod Kumar Srivastava, PhD, Swati Verma, MPharm

Other topics/volumes are planned on these topics:

1. Globalization and Emerging Leadership
2. Positive Psychology and Today's Organizations
3. Changing Consumer Behavior and Organizations
4. The Impact of Technological Advancement on Organizations
5. Emerging Employer and Employee Relations
6. Designing Future Organizations and Emerging Sectors
7. Aging in South Asia and the Impact on Emerging Businesses
8. Issues in Intercultural Management
9. The Role of Spirituality in Management
10. Increasing Workforce Diversity in Organizations
11. Creating Innovation in Organizations
12. Purchasing Power and Happiness in Customers

ABOUT THE BOOK SERIES EDITOR

Arvind K. Birdie, PhD, has been with various organizations in different roles and has more than eighteen years of experience in higher education. Presently she is an Associate Professor in the School of Humanities, Social Sciences and Education at G. D. Goenka University, Sohna, Gurgaon, India. She was formerly Assistant Director of International Affairs at Amity University, Gurgaon. As an avid reader, she has expertise in teaching various interdisciplinary subjects with equal ease. In addition to academic teaching and training, Dr. Birdie organizes management development programs for corporate and academicians. She is a regular presenter at international and national conferences, and she has published papers in refereed journals. Her areas of interest include leadership, work-life balance, virtual work, and positive psychology. Dr. Birdie has been honored with the Prof. Mrs. Manju Thakur Memorial Award 2016 for Innovative Contributions in Research/Test Construction/Book Publication for her book Organizational Behavior and Virtual Work: Concepts and Analytical Approaches, presented at the 52nd National and 21st International Conference of the Indian Academy of Applied Psychology. She is also Editor-In-Chief of the book series 21st Business Management, published by Apple Academic Press.

About the Editors

Arvind K. Birdie, PhD
Associate Professor, School of Humanities, Social Sciences and Education, G. D. Goenka University, Sohna, Gurgaon, India

Arvind K. Birdie, PhD, is an enthusiastic, versatile academician and learner to the core, with more than 18 years of experience in higher education. She has been with various organizations in different roles. Presently she is associated with G. D Goenka University, Gurgaon, India, as an Associate Professor and PhD Program Leader. In the past, she was Asst. Director, International Affairs at Amity University, Gurgaon. As an avid reader, she has expertise in teaching various interdisciplinary subjects with equal ease. In addition to academic teaching and training, she has also organized management development programs for corporations and academicians. She is a regular presenter at various international and national conferences and has published papers in refereed journals. Her areas of interest include leadership, work-life balance, virtual work, and positive psychology. Dr. Birdie has been honored with the Prof. Mrs. Manju Thakur Memorial Award 2016 for Innovative Contributions to Research/Test Construction/Book Publication for her book *Organizational Behavior and Virtual Work: Concepts and Analytical Approaches*. The award was presented during the 52nd National and 21st International Conference of the Indian Academy of Applied Psychology, held at the Department of Psychology, University of Rajasthan, Jaipur, February 23rd–25th, 2017. She is an Editor in Chief for 21st Business Management Series for Apple Academic Press (CRC/Taylor & Francis).

Ruchi Joshi, PhD
Assistant Professor, Department of Psychology,
Manipal University Jaipur, India

Ruchi Joshi, PhD, is presently working as an Assistant Professor in the Department of Psychology at Manipal University Jaipur, India. She has rich experience of over a decade in teaching, research, and counseling in different organizations and has been associated with reputed organizations such as Mohan Lal Sukhadia University Udaipur, India, and Amity University Rajasthan, India. She has a number of national and international publications to her credit that includes publications which are peer-reviewed, UGC CARE, and Taylor and Francis-indexed. She has also presented papers at a number of international, national, and regional conferences, seminars, and workshops. In addition to teaching, she also shares her expertise as a resource person and panelist at various platforms, such as workshops, panel discus-sions, seminars, and webinars on issues related to general well-being, mental health, physical, and sexual violence.

Contents

Contributors

Chirmi Acharya
Assistant Professor, Department of Psychology, Manipal University Jaipur, Rajasthan, India

Samo Bobek
University of Maribor, Faculty of Economics and Business, Slovenia

Zdenko Deželak
University of Maribor, Faculty of Economics and Business, Slovenia

Neeta Gupta
Associate Professor, DAVPG College, Dehradun, Uttarakhand, India

Monika Gwalani
Assistant Professor, Amity University, Jaipur, Rajasthan, India

Ruchi Joshi
Assistant Professor, Department of Psychology, Manipal University Jaipur, India

Kashish Kakra
IND Analyst II-HCS, Aon Consulting Pvt. Ltd., Gurgaon, India

Attin Kumar
Cluster Manager, Reliance Trends, Gurgaon, India

Nida Nafees
Assistant Professor, Amity Institute of Behavioral and Allied Sciences, Amity University, Rajasthan, India

Vismita Paliwal
Associate Professor and Coordinator, Amity Institute of Behavioral and Allied Sciences, Amity University, Rajasthan, India

Anita Peša
University of Zadar, Department of Economics, Croatia

Mladen Rajko
University of Zadar, Department of Economics, Croatia

Rimjhim
Mphil Clinical Psychology, Department of Psychiatry, Government Medical College and Hospital, Chandigarh

Maja Rožman
University of Maribor, Faculty of Economics and Business, Slovenia

Tarun Kumar Sharma
Assistant Professor, Department of Psychology, Mohanlal Sukhadia University, Udaipur, Rajasthan, India

Varsha Sharma
Assistant Professor, Department of Psychology, Mohanlal Sukhadia University, Udaipur, Rajasthan

Suyesha Singh
Assistant Professor, Department of Psychology, Manipal University Jaipur, Rajasthan, India

Neharshi Srivastava
Assistant Professor, Amity Institute of Behavioral and Allied Sciences, Amity University, Jaipur, Rajasthan, India

Tjaša Štrukelj
University of Maribor, Faculty of Economics and Business, Slovenia

Priya Vij
Professor, Economics, Chandigarh University, Punjab, India

Simona Sternad Zabukovšek
University of Maribor, Faculty of Economics and Business, Slovenia

Abbreviations

AI	artificial intelligence
AVE	average variance extracted
CJRS	coronavirus job retention scheme
COVID-19	coronavirus infection 2019
CR	composite reliability
CSA	customer service associate
ETR	enterprise technology research
FMCG	fast-moving consumer goods
FOMO	fear of missing out
IAT	internet addiction test
MAP	museum of art and photography
NASSCOM	National Association of Software and Services Companies
NCPA	National Center for Performing Arts
NFI	normed fit index
NIOSH	National Institute of Safety and Health
NIS	nursing incivility scale
OTWI	Organizational tolerance Workplace Incivility
PHE	public health emergency
PPE	personal protective equipment
RT	reality therapy
SARS-CoV-2 Virus	severe acute respiratory syndrome corona virus
SBCs	session border controllers
SMT	self-management training
SPSS	statistical package for the social sciences
SRMR	standardized root mean square residual
TAT	turnaround time
TCS	Tata Consultancy Services
UC	unified communications
UCaaS	unified communications as a service
UI	user interface
WFH	work from home

Preface

November 2019 marked the first ever case of the novel virus COVID-19, or what is commonly known as coronavirus in China, and little did the world know that it would engulf every continent, and every country bringing the world to its knees, stopping all activity, halting all movement.

Not only poor and under-developed countries, but even developed countries and so-called superpowers saw hospital beds occupied, ventilators, protective gear, masks, and medicines falling short with medical fraternity perplexed and overburdened and political leaderships clueless.

That the pandemic changed the way of life would be an understatement. It would be more appropriate to say that the pandemic made us adapt to a new way of life. Barely a few days into the pandemic and we had already coined new terms like "new normal," "post-pandemic," etc.

The pandemic cost numerous lives and millions took to hospitalization and underwent treatment across the globe. Lockdowns that were imposed worldwide to check the spread of infection did flatten the infection curve but it unfortunately also led to flattening the curve of economic growth.

Resilience is one of the most impactful driving forces in human life. Despite extreme adversities and trauma, we initially crawl back but then stand tall and reach new milestones. And this is exactly what happened when the will power to overcome the chaos, confusion, and turbulence caused by the pandemic led to new adapting and coping strategies. We made unheard advancement in the field of research, medicine, and technology and went back to what was termed as the "new normal." This invincible streak to win was reflected in the economic front as well.

The present book is an effort to map how the world has changed not only in terms of work or business life but also other linking areas of life with adaptation towards post-pandemic effects. Areas and avenues in all walks of the business world which are diverse but indirectly connects with each other are covered in the form of 10 chapters through out this volume (4th book) under the 21st Business Management Series.

With the help of multidimensional model, Chapter 1 tries to explore the COVID-19 impact on well-being and aims to understand leadership with employee satisfaction, organization's future competitiveness and quality of life.

Chapter 2 explains how organizations opted and sustained the Covid-19 pandemic through work from home (WFH) and how organizations started exploring the hybrid working style as an alternative option which we will see as the future of organizations. The obvious challenges and complications of flexible working are discussed with emphasis on developing hybrid work organization policies.

The discussion on the workplace incivility in recent post-COVID times and how organization culture affects the workplace making it uninhabitable at times with the help of metanalysis of literature review and antecedents are mentioned in Chapter 3. Suggestions for actions to promote a positive workplace are also analyzed.

The collaboration platforms have become an important part of work life in recent years. Through Chapter 4, authors examine this as due to COVID the changes in the user experience of collaboration platforms such as those of Microsoft teams have been analyzed. Digital collaboration, user experience (UX) are some of the concepts are discussed in this chapter.

Not only serious and monotonous business but entertainment industry faced the heat as well. Art, culture, cinema, and creativity business are also significant revenue generators. The effect of the pandemic on the lives and earnings of artists was studied in qualitative study through chapter 5. Chapter 5 attempts to explore the impact of pandemic on social and psychological aspects in art and culture industry and how governments and organizations can assist in reviving the artists as well as the whole industry.

Chapter 6 explores important aspect for every employee, especially those working virtually during the pandemic and that is psychological well-being. It tries to investigate whether resilience proves to be a predictor of psychological well-being among e-workers amid pandemic. Recommendations and interventions for improving psychological well-being are also highlighted.

Food is the very essence of life and in a way the reason why every economic activity is undertaken. Pandemic had devastating and far-reaching effects on food, beverages, and travel industry. If the world-renowned hotels, restaurants, cafe, and eateries chains experienced a plunge in earnings and struggled hard to thrive, how the small and unorganized food industry was more pressed and struggled to survive is documented in Chapter 7 based on a qualitative study.

Due to the pandemic, many employees working in organizations all of a sudden had to work online, and many lost their jobs and businesses. The stress of work and life goes hand-in-hand. The new work-life balance and new normal has been studied under Chapter 8. Effective strategies for this new normal and work-life balance are also discussed in this chapter.

Chapter 9 evaluates the internet addiction and mental health challenges which emerged due to constant working and using digital gadgets for red collar employees. The chapter tries to explore the effect of Group Therapies (Autogenic Relaxation Training and Self-Management Training) for reducing internet addiction among Red Collar employees after the pandemic.

The world during COVID-19 pandemic witnessed massive changes in consumers' preferences and businesses are responding to sustain their margins. The 21st century saw a digital transformation where everything is available online, from groceries to gadgets, the outbreak of COVID-19 just speeded up this transformation. Chapter 10 discusses how consumer choices and shopping psychology has changed due to the pandemic, especially FMCG sector. The chapter also discusses the policy implication with respect to consumers and businesses' adaptation to the new normal.

This is the fourth book in the book series 21st Century Business Management, which aims to provide insights and changes in business management.

The approach in developing chapters is by reviewing and conducting empirical studies, qualitative analysis, round tables and focus discussion with employees and employers of service industry by the authors.

We extend a generous gratitude towards all the contributors without whose determined efforts this work would not have seen the light of the day.

Their rich insights and shared perspectives have enabled them to bring in a lot of unique concepts of interdisciplinary ways to understand today's business world. We are indebted to organizations and a lot of industry practitioners who provided support and discussed their ideas with the authors. We express special thanks to the whole team of Apple Academic Publishers for their constant and unquestionable support without which this work would not have been reality.

In a nutshell, this book is a tribute to the indomitable will to overcome every adversity, an effort to document the legacy of hope and optimism for future generations and last but not the least, rather most importantly, a prayer for all the near and dear ones who were lost to the pandemic and an encouraging tap on the shoulders of those who moved on with tears in their eyes and memories in their hearts.

CHAPTER 1

COVID-19 and Employee Satisfaction with Leadership Influencing the Future Competitiveness of Organizations

MAJA ROŽMAN,[1] ANITA PEŠA,[2] MLADEN RAJKO,[2] and TJAŠA ŠTRUKELJ[1]

[1]*University of Maribor, Faculty of Economics and Business, Slovenia*

[2]*University of Zadar, Department of Economics, Croatia*

ABSTRACT

The COVID-19 pandemic has a strong influence on daily lives, health, interpersonal, society, economic, financial, and ecological fields. The COVID-19 outbreak and containment measures have a long-term impact on the economy, businesses, and the working environment, thus, all individuals, society, and planet Earth. During COVID-19, employees must face new ways of working and adapt to changing circumstances in their private lives. The situation caused by the COVID-19 crisis will also affect our future. The old ways of our work and life in the new reality of the changing social patterns will no longer be possible. Many organizations are exploring working from home as a temporary or alternative working arrangement. Still, when employees perform their work at home, they face different problems than when they work in the workplace, leading to differences in their satisfaction with working conditions, well-being, work efficiency, and work engagement. Therefore, organizations should be aware of the importance of an appropriate work environment and appropriate leadership expressed through employer care for employees

Reshaping the Business World Post-COVID-19: Management Strategies for Sustainable Behavior Change. Arvind K. Birdie & Ruchi Joshi, (Eds.)

who perform work from home (WFH) during the COVID-19 pandemic. The chapter aims to analyze the effects of employee satisfaction with leadership, which is influencing an organization's future competitiveness, through five constructs: (i) employee satisfaction with working conditions; (ii) employee satisfaction with employer care about their well-being; (iii) employee satisfaction with employer care about their work engagement; (iv) employee satisfaction with employer care about their work efficiency; and (v) employee satisfaction with employer care about their work motivation during the COVID-19 pandemic. The chapter also aims to develop the multidimensional model and analyze the strength of the relationships among constructs in the model. The research was implemented on a sample of 885 employees. Based on the research, we can confirm positive relationships in: (a) employee satisfaction with working conditions; (b) employee satisfaction with employer care about their well-being; (c) employee satisfaction with employer care about their work efficiency; (d) employee satisfaction with employer care about their work engagement; and (e) employee satisfaction with employer care about their work motivation among employees during the COVID-19 pandemic. Based on the research, we can confirm that although there are differences in the strength of the relationships among researched constructs in the multidimensional model measuring employees' satisfaction with leadership during the COVID-19 pandemic, there is evidence that more satisfied employees, employees who perceive employer care: (1) feel well; (2) are more motivated; (3) work engaged; and thus (4) efficient; and therefore (5) more contribute to the organization's future competitiveness.

1.1 INTRODUCTION

The COVID-19 pandemic created many disruptions and chaos that have affected many millions of lives worldwide and changed businesses. Organizations witnessed one of the COVID-19 eras, the worst crises in history, and are struggling to survive. Thus, the COVID-19 pandemic posed a series of challenges and dilemmas (Adikaram et al., 2021). The COVID-19 pandemic and the consequent lockdown caused significant economic disruption, leading organizations to make a rapid transition and take a new approach to business strategy (Diab-Bahman & Al-Enzi, 2020). To maintain a competitive position, organizations constantly adapt work processes in their management and basic realization processes. Their exceptional attention is often focused on training their employees and devoting much more time to communicate with employees (He et al., 2020). As a result, during COVID-19, organizations face reduced demand for products, a substantial decline in investment, and lower performance (Shen et al., 2020).

In addition, organizations are overburdened with ensuring their employees' health, well-being, and safety and limiting the spread of the virus, leading to drastic changes in work settings and work arrangements (Kumar et al., 2020). These crises have caused mental health problems among employees due to social isolation, reduced work efficiency, and increased employee uncertainty about their health, job, and futures (Rudolph et al., 2020). However, global pandemics and economic crises do not affect all employees equally. Most employees have followed strict and often uncomfortable measures, and prolonged exposure to those can lead to mental distress or overwork (Amoah & Simpeh, 2020). The COVID-19 global health emergency and its economic and social impacts have disrupted nearly all aspects of life for all groups in society (Toscano & Zappalà, 2020). In this new environment, employers have to adapt and make contingency plans to respond to new measures as they arise. Employees are experiencing the effects of COVID-19 in different ways, but they are affected by this new situation (Madero Gómez et al., 2020). Many organizations have introduced new ways of working, such as teleworking (working from home) as a temporary or alternative working arrangement to continue their business operations during COVID-19 (Kapoor et al., 2021). According to Nguyen (2020), organizations inexperienced in working from home practices learn to lead, engage, satisfy, motivate, and interact with their employees in new ways (Nguyen, 2020). New ways of working bring significant changes to employees, and employees face different problems than when they work in the workplace (Diab-Bahman & Al-Enzi, 2020).

For example, many parents have to harmonize somehow work commitments and childcare or housework (Anderson & Kelliher, 2020). During the lockdown, childcare was a significant cause for concern for employees who are parents and poses the biggest challenge to labor within the home (Chung, 2020). In addition, the new way of working from home (WFH) presents challenges of finding suitable space for working, access to equipment, and reliability of internet connectivity. Also, employees face other stresses (for example, caring for the elderly) and challenges of working from home during the COVID-19 pandemic (Anderson & Kelliher, 2020). The outbreak of COVID-19 has generated more stress and poor mental health among employee parents in the United States (Hamel & Salganicoff, 2020).

Moreover, during the lockdown, childcare and other support services have not been available for many working parents, increasing pressure on working parents and limiting their work efficiency (Uddin, 2020). Also, other employees, because of isolation, increased care for the elderly, and poor leadership in organizations, face stress, depression, and anxiety (Kapoor et al., 2021; Madero Gómez et al., 2020). Therefore, organizations should be aware of the importance of appropriate leadership expressed through employer care for employees who perform WFH during the COVID-19 pandemic; additionally, employee satisfaction with leadership influences an organization's future competitiveness.

1.2 LITERATURE REVIEW

1.2.1 EMPLOYEE SATISFACTION WITH WORKING CONDITIONS DURING THE COVID-19 PANDEMIC

Employee satisfaction influences not only employee efficiency but also employee performance. It also influences how an organization's goals are achieved to improve customer satisfaction, customer loyalty and satisfaction, and perceived service quality (Díaz-Carriónb et al., 2020). Teleworking (working from home) was not an option for all employees in different industries before COVID-19; therefore, the COVID-19 pandemic has brought new aspects of working conditions to which employees must adapt (Bhattarai, 2020; Diab-Bahman & Al-Enzi, 2020). Employees: (i) stress due to familiarity with technology that is rarely or never used; (ii) getting used to the paperless administration and guidelines; (iii) feeling of insecurity; (iv) fear of infection; (v) managing office space at home; and (vi) achieving work-life balance affect work efficiency among employees who WFH during the COVID-19 pandemic (Bhattarai, 2020). Nagel (2020) emphasizes that teleworking increases employees' autonomy who WFH during COVID-19, leading to a higher level of employee satisfaction with working conditions (Nagele, 2020). Employees who WFH during

COVID-19 have higher autonomy. Still, on the other hand, work-family conflict, a poor balance between work and private life, and care for the elderly or children increase occupational stress and decrease work efficiency (Amoah & Simpeh, 2020).

Gavidia (2020) found out that during COVID-19, occupational stress is higher among employees (who WFH) compared to the period before COVID-19. Results showed that 88% of employees reported stress over the past 4 to 6 weeks. Among those reporting stress, 62% noted losing at least 1 hour a day in productivity, and 32% lost at least 2 hours a day in productivity (Gavidia, 2020). The employee satisfaction with working conditions that influence employee satisfaction with employer care about their well-being depends on employee satisfaction with leadership during COVID-19 (Mgammal & Al-Matari, 2021). According to Dalkrani & Dimitriadis (2018), positive or negative employee satisfaction with working conditions highly results from different factors, such as salary, job security, promotional possibilities, and the working climate (Dalkrani & Dimitriadis, 2018). When employees are satisfied with working conditions, this has a positive effect on employee satisfaction with employers caring about their well-being because they feel that the employer is committed to caring for their health and well-being (Davidescu et al., 2020). Also, because employees who WFH have less personal contact with other employees and people, therefore these employees should be at lower risk of infection. This decreased risk could lead to an impression of being looked out for and valued and increase feelings of security, thus leading to greater employee satisfaction (Nagele, 2020). Additionally, when employees are satisfied with working conditions, this has a positive effect on employee satisfaction with employer care about their work efficiency because they feel that the employer is supportive, valuing, and caring for their employees (He et al., 2020; Kapoor et al., 2021).

Hence, it is proposed:

- **H1:** *Employee satisfaction with working conditions has a positive direct effect on employee satisfaction with employer care about their work efficiency, among employees who work from home during COVID-19.*
- **H2:** *Employee satisfaction with working conditions has a positive direct effect on employee satisfaction with employer care about their well-being among employees who work from home during COVID-19.*

1.2.2 EMPLOYEE SATISFACTION WITH EMPLOYER CARE ABOUT THEIR WELL-BEING DURING THE COVID-19 PANDEMIC

Employee well-being during COVID-19 has been recognized as a significant issue for employers (Meyer et al., 2021). Organizations with appropriate leadership can achieve a competitive advantage during COVID-19 with employees who feel good (Tuzovic & Kabadayi, 2020). Employee well-being is linked to various working conditions, which increase or decrease occupational stress and work-life balance among employees (Keeman et al., 2017). Magnavita et al. (2020) found that emotional and behavioral symptoms of health problems among employees during the COVID-19 pandemic increase. They found that there was a significant increase in anxiety, depression, and insomnia. Also, Magnavita et al. (2020) found that physical symptoms among employees during COVID-19 increased, reflecting low sleep quality and stomachaches. Additionally, Meyer et al. (2021) found out that employees who WFH during the COVID-19 pandemic face increasing exhaustion and headaches.

On the other hand, Lee (2021) emphasizes that appropriate leadership during COVID-19 plays an essential role in employee well-being, and organizations that have introduced appropriate working conditions report about better employee well-being during COVID-19 (Lee, 2021). Employees value when the organization takes care of their well-being, and also, employees value support, which the organization gives during COVID-19 (Anderson & Kelliher, 2020). High support from the organization manifested within a healthy and positive interpersonal relationship with employer and employees fosters high psychological safety and well-being at work (Baert et al., 2020). From this point of view, employees are satisfied with the employer's care about their well-being leading to employee satisfaction with employer care about their work engagement (Wong et al., 2020). When employees feel happy in performing their work, this leads to higher work engagement (Joo & Lee, 2017). Employee well-being depends on satisfaction with leadership and work conditions which lead to higher employee satisfaction with employer care about their work engagement among employees who WFH during the COVID-19 (Abbas & Zhiqiang, 2020; Chanana & Sangeeta, 2020; Davidescu et al., 2020).

According to this, the following hypothesis is proposed:

- H3: *Employee satisfaction with employer cares about their well-being has a positive direct effect on employee satisfaction with employer care about their work engagement among employees who work from home during the COVID-19.*

1.2.3 EMPLOYEE SATISFACTION WITH EMPLOYER CARE ABOUT THEIR WORK ENGAGEMENT DURING THE COVID-19 PANDEMIC

During COVID-19, new work settings and work arrangements change the work-life balance among employees who WFH (Madero Gómez et al., 2020). Teleworking has allowed employees to do their work and their personal needs simultaneously (for example, taking care of children, the elderly, and other errands). The problem is that many non-work-related factors can harm an employee's mindset and, therefore, their work engagement (Adikaram et al., 2021; Toscano & Zappalà, 2020). In addition, many aspects of teleworking can negatively affect work engagement, such as organizational politics, administrative hassles, a lack of resources, role conflict, and excessive workloads (Kapoor et al., 2021; Madero Gómez et al., 2020). Millard & Blackburn (2020), in their research on 282 employees, found out that they face higher anxiety and less work engagement during the COVID-19, which leads to a low level of employee efficiency among employees who WFH (Millard & Blackburn, 2020). During COVID-19, appropriate leadership is an essential viewpoint because it positively impacts work engagement, mainly through support and feedback (Haque, 2021; Abbas & Zhiqiang, 2020). Regular contact between employees and leaders is vital to ensure employees receive sufficient support and feel valued (Caringal-Go et al., 2021). During COVID-19, employees need more understanding from the leader about their work, and also, employees must feel supported, connected, and recognized (Bhattarai, 2020). The leader's critical role is to support employees by creating a safe space for concerns around mental, emotional, and physical health and ensuring employees feel connected to their organization even amidst a shift to remote work (Davidescu et al., 2020; Chanana & Sangeeta, 2020). Thus, employee satisfaction with employer care about their work engagement leads to higher employee satisfaction with employer care about their work efficiency among employees who WFH during COVID-19 (Dwivedi et al., 2020; Rudolph et al., 2020).

Therefore, the following hypothesis is proposed:

- H4: *Employee satisfaction with employer care about their work engagement has a positive direct effect on employee satisfaction with employer care about their work efficiency among employees who work from home during the COVID-19.*

1.2.4 EMPLOYEE SATISFACTION WITH EMPLOYER CARE ABOUT THEIR WORK MOTIVATION DURING THE COVID-19 PANDEMIC

Tovmasyan & Minasyan (2020), in their research on 100 employees, found that employees are most motivated when they perform their work in the workplace, and for only 12% of respondents, remote working is motivating. Thus, results show that in Armenia, employees prefer to work from the workplace. Also, Gensler's US Work from Home Survey (2020) on 2,300+ full-time US employees across 10 different industries found that only 12% of employees want to WFH full-time. Most employees (70%) want to spend most of the workweek at the office while working from home when they wish. About 30% want a flexible work arrangement. Results of the research show that 26% of employees want to WFH for 1 or 2 days a week, and 18% of employees want to WFH for 3 or 4 days a week. Also, Gensler's US Work from Home Survey (2020) analyzes the reasons why employees want to come back to the workplace. The reasons are scheduled meetings with colleagues (54%), socializing with colleagues (54%), impromptu face-to-face time (54%), being part of the community (45%), access to technology (44%), focusing on work (40%), scheduled meetings with clients (40%), professional development/coaching (33%) and access to amenities (29%). Therefore, Wiradadi Color et al. (2020) emphasize a challenge for the organization and leadership on maintaining employee motivation during the COVID-19 pandemic. A higher level of anxiety among employees during COVID-19 will negatively affect employee motivation which affects the level of negative emotions. This lead to a lower level of work engagement and work efficiency (Wiradadi Color et al., 2020). Employees will be motivated to work and perform their work well if their safety is thought of by the organization (Mensah & Tawiah, 2016). Organizations that make decisions and policies to provide security for employees are essential to keep growing positive motivation from employees, especially during the COVID-19 pandemic (Wiradadi

Color et al., 2020). This also leads to higher employee satisfaction with employer care about their work motivation: because the organization becomes a central role in increasing employee motivation with motivation factors and with a sense of security for employees, which have a positive effect on work engagement and work efficiency among employees during the COVID-19 (Wiradadi et al., 2020; Millard & Blackburn, 2020; Tovmasyan & Minasyan, 2020).

According to this, the following two hypotheses are proposed:

- **H5:** *Employee satisfaction with employer cares about their work motivation has a positive direct effect on employee satisfaction with employer care about their work engagement, among employees who work from home during the COVID-19.*
- **H6:** *Employee satisfaction with employer care about their work motivation has a positive direct effect on employee satisfaction with employer care about their work efficiency, among employees who work from home during COVID-19.*

1.2.5 EMPLOYEE SATISFACTION WITH EMPLOYER CARE ABOUT THEIR WORK EFFICIENCY DURING THE COVID-19 PANDEMIC

Social isolation and occupational stress among employees, who WFH during COVID-19, lead to lower work efficiency (Toscano & Zappalà, 2020). Also, Toscano & Zappalà (2020), in their research on 265 employees, found that in teleworking during the COVID-19 pandemic, the social isolation generated by the lack of face-to-face contact with employees is positively associated with occupational stress, which is reflected in lower work efficiency. Employees need support and exemplary leadership during this challenging time (Dwivedi et al., 2020). During COVID-19, employees often face difficulties how to balance all roles and responsibilities, especially those related to work and family (Žnidaršič & Bernik, 2021). Work-family balance is already becoming a social and institutional problem facing all countries. Organizations still do not recognize the importance of reconciling work and family life during COVID-19 (Kapoor et al., 2021). Lapowsky's (2020) research shows that women employees with children at home reported more stress during the

COVID-19 pandemic than women without children (46% compared to 34%). In addition, women employees with children at home face lower work efficiency. The results show the same for men employees with children at home – 71% of men with children at home reported higher stress levels and lower work efficiency during the COVID-19 pandemic. The organization should support employees in work-family balance with various family-friendly practices and policies during COVID-19 (e.g., flexible working hours, part-time work, "compressed" work week) (Dwivedi et al., 2020). Also, leadership support is essential because employee satisfaction with the employer cares about their work efficiency during the COVID-19 influencing organization's future competitiveness (Haque, 2021; Dwivedi et al., 2020). The perception of organizational support for employees reduces conflicts between work and family and positively impacts achieving successful results (Žnidaršič & Bernik, 2021). According to the above hypothesis was proposed the conceptual model presented in Figure 1.1.

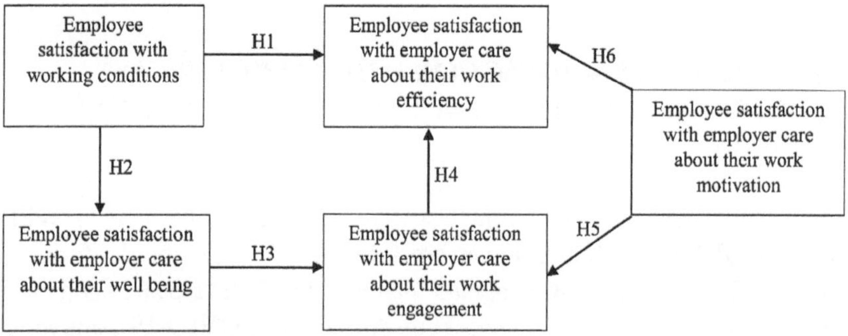

FIGURE 1.1 Conceptual model of research.

1.3 METHODOLOGY

1.3.1 PARTICIPANTS

The research, as part of a broader research, involved randomly selected 177 organizations in Slovenia, and from each organization, five employees participated in our research. Thus, 885 employees responded to the questionnaire. In addition, the employer randomly selected five employees who responded to the questionnaire.

The research during the COVID-19 pandemic involved 11.0% of employees aged from 26 to 31 years, 13.2% of employees aged from 32 to 37 years, 16.0% of employees aged from 38 to 43 years, 21.4% of employees aged from 44 to 49, 25.4% employees aged from 50 to 55 years, 7.6% employees aged from 56 to 61 years. The lowest percentage is presented by employees aged over 62 (5.4%) years. The organizations in which employees are employed were from: (i) manufacturing (22.0%); (ii) trade, maintenance, and repair of motor vehicles (16.4%); (iii) financial and insurance activities (14.1%); (iv) information and communication activities (10.3%); (v) professional, scientific, and technical activities (10.7%); (vi) real estate business (12.4%); (vii) other diversified business activities (7.3%); (viii) catering (2.8%); (ix) transport and storage (2.3%); and (x) construction (1.7%). The most significant share of organizations in which employees are employed present medium-sized organizations (40.7%). Large organizations present a 35.6% employees sample, and small organizations present a 23.7% employees sample.

1.3.2 MEASURES

For the research instrument, we used a questionnaire – closed type: (i) items for the construct of employee satisfaction with working conditions were adopted from O'Moore & Lynch (2007); (ii) items for employee satisfaction with employer care about their well-being were adopted from Jaiswal & Dyaram (2019); (iii) items for employee satisfaction with employer care about their work engagement were adopted from Robertson & Cooper (2010); (iv) items for the construct employee satisfaction with employer care about their work efficiency were adopted from Dwivedi et al. (2020); (v) items for the construct employee satisfaction with employer care about their work motivation were adopted from Conrad et al. (2015). The respondents indicated their agreement to the listed statements at a 5-point Likert-type scale from 1-disagree entirely to 5-agree entirely.

1.3.3 DATA ANALYSIS

In the first part, we used descriptive statistics to show the average rate for each statement and standard deviation, which measures the dispersion of a dataset relative to its mean.

In the second step, structural equation modeling with PLS-SEM was used. PLS-SEM can relate the set of independent variables to multiple dependent variables, and it can also handle many independent variables even when predictors display multicollinearity (Hair et al., 2016). In the first phase of data processing, the confirmatory factor analysis was used to: (i) test the dimensionality of constructs; and (ii) establish discriminant validity and convergent validity. The reliability of measurement scales was assessed within the scope of inner consistency with Cronbach's alpha coefficient (Chronbach, 1951). The average variance extracted (AVE) was used for testing convergent validity, and the composite reliability (CR) measure was used for testing reliability. Discriminant validity was assessed with a criterion comparing the square root of AVE to all correlation coefficients (Fornell & Larcker, 1981). The normed fit index (NFI) and standardized root mean square residual index (SRMR) fit indices were used to assess the fit of the measurement model. Statistical package for the social sciences (SPSS) and SmartPLS 3 software tool was used for analyzing the data.

1.4 RESULTS

1.4.1 THE RESULTS OF DESCRIPTIVE STATISTICS FOR EACH CONSTRUCT

In the beginning, Tables 1.1–1.5 show the results of descriptive statistics for each individual statement for five constructs, which are: (i) employee satisfaction with working conditions; (ii) employee satisfaction with employer care about their well-being; (iii) employee satisfaction with employer care about their work engagement; (iv) employee satisfaction with employer care about their work efficiency; and (v) employee satisfaction with employer care about their work motivation.

The results in Table 1.1 shows that the means for answers about employee satisfaction with working conditions indicate that, on average, employees who WFH during the COVID-19 had the highest agreement with: I am satisfied with the flexible forms of work that the organization allows me during the COVID-19 (mean: 3.80), I am satisfied with enabling flexible working time (mean: 3.78), I am satisfied with the number of breaks during work which I perform from home (mean: 3.77),

TABLE 1.1 Descriptive Statistics for Construct Employee Satisfaction with Working Conditions

Statement	Mean	Median	Standard Deviation
I am satisfied with the internal communication.	3.71	4.00	1.120
I am satisfied with all the information regarding the changes in the organization.	3.73	4.00	1.144
I am satisfied with enabling flexible working time.	3.78	4.00	1.124
I am satisfied with the balance between work and private life.	3.58	4.00	1.234
I am satisfied with the daily schedule of work obligations.	3.69	4.00	1.216
I am satisfied with the number of different training programs (online).	3.53	4.00	1.177
I am satisfied with teleworking (work from home).	3.64	4.00	1.335
I am satisfied with the number of breaks during work which I perform from home.	3.77	4.00	1.174
I am satisfied with the flexible forms of work that the organization allows me during COVID-19.	3.80	4.00	1.418
I am satisfied with the leadership.	3.74	4.00	1.468

I am satisfied with the leadership (mean: 3.74), I am satisfied with all the information regarding the changes in the organization (mean: 3.73), I am satisfied with the internal communication (mean: 3.71), I am satisfied with the daily schedule of work obligations (mean: 3.69), I am satisfied with teleworking (WFH) (mean: 3.64), I am satisfied with the balance between work and private life (mean: 3.58), I am satisfied with the number of different training programs (online) (mean: 3.53). The highest dispersion (standard deviation) of employees who WFH during COVID-19 is observed in satisfaction with the leadership. This means that agreement with the statements differs most in this indicator. On the other hand, the lowest dispersion of employees, who WFH during COVID-19, is observed in satisfaction with internal communication.

Table 1.2 shows that the means for answers about employee satisfaction with employer care about their well-being indicate that, on average, employees who WFH during the COVID-19 had the highest agreement with: When I have work problems, the employer takes time to listen to me (mean: 3.73), The leader is always ready to help me when I have problems (mean: 3.72), the employer cares that I feel good when I perform WFH

TABLE 1.2 Descriptive Statistics for Construct Employee Satisfaction with Employer Care About Their Well-Being

Statement	Mean	Median	Standard Deviation
When I have work problems, the leader takes time to listen to me.	3.73	4.00	1.325
The employer is always ready to help me when I have problems.	3.72	4.00	1.349
The employer cares that I feel good when I perform work from home.	3.64	4.00	1.321
The employer cares that I am not subject to work stress.	3.61	4.00	1.293
The employer cares that I do not feel workload.	3.58	4.00	1.442
The employer cares that I do not feel the behavioral symptoms of burnout.	3.45	4.00	1.501
The employer cares that I do not feel the physical symptoms of burnout.	3.41	5.00	1.797
The employer cares that I do not feel the emotional symptoms of burnout.	3.39	5.00	1.803
The employer organizes different seminars about health, well-being, stress, etc.	3.35	5.00	1.815

(mean: 3.64), the employer cares that I am not subject to work stress (mean: 3.61), the employer cares that I do not feel workload (mean: 3.58). On the other hand, results indicate that employees who WFH during the COVID-19, on average, had partial agreement with: The employer cares that I do not feel the behavioral symptoms of burnout (mean: 3.45), the employer cares that I do not feel the physical symptoms of burnout (mean: 3.41), the employer cares that I do not feel the emotional symptoms of burnout (mean: 3.39), the employer organizes different seminars about health, well-being, stress, etc. (mean: 3.35). The highest dispersion of employees who WFH during the COVID-19 is observed in the statement: "The employer organizes different seminars about health, well-being, stress, etc.," and the lowest dispersion of employees who WFH during the COVID-19 is observed in the statement: "The employer cares that I am not subject to work stress."

Table 1.3 shows that the means for answers about the employee satisfaction with employer care about their work engagement indicate that, on average, employees who WFH during the COVID-19 had the highest agreement with: The employer cares that I have everything I need to do

TABLE 1.3 Descriptive Statistics for Construct Employee Satisfaction with Employer Care About Their Work Engagement

Statement	Mean	Median	Standard Deviation
The employer cares that I do my work with passion when I perform work from home during COVID-19.	3.63	5.00	1.785
The employer cares that I am engaged in the quality of my work during COVID-19.	3.78	5.00	1.676
The employer cares that I am engaged in achieving successful business results during COVID-19.	3.70	5.00	1.779
The employer cares that I am aware of the importance of innovation for our organization and that I am helping to develop the organization during COVID-19.	3.61	4.00	1.419
The employer cares that I feel a connection with the organization in which I worked.	3.67	4.00	1.372
The employer cares that I feel that my work and job are essential.	3.68	4.00	1.359
The employer takes my business ideas into account.	3.72	4.00	1.188
The employer cares that I have everything I need to do my job well during COVID-19.	3.86	4.00	1.204
Over the last year, I had the opportunity to develop.	3.45	4.00	1.431

my job well during the COVID-19 (mean: 3.86), the employer cares that I am engaged in the quality of my work during the COVID-19 (mean: 3.78), the leader takes my business ideas into account (mean: 3.72), the employer cares that I am engaged to achieve successful business results during the COVID-19 (mean: 3.70), the employer cares that I feel that my work and job are essential (mean: 3.68), the employer cares that I feel connection with the organization in which I worked (mean: 3.67), the employer cares that I do my work with passion when I perform WFH during the COVID-19 (mean: 3.63), the employer cares that I am aware of the importance of innovation for our organization and that I am helping to develop the organization during the COVID-19 (mean: 3.61). On the other hand, employees partly agree with the statement: Over the last year, I had the opportunity to develop (mean: 3.45). The highest dispersion of employees who WFH during COVID-19 has been observed in the statement: "The employer cares that I do my work with passion when I perform WFH during COVID-19." On the other hand, the lowest dispersion of employees who WFH during COVID-19 has been observed in the statement: "The leader takes my business ideas into account."

TABLE 1.4 Descriptive Statistics for Construct Employee Satisfaction with Employer Care About Their Work Efficiency

Statement	Mean	Median	Standard Deviation
The employer cares that my workability has not decreased during COVID-19.	3.68	5.00	1.673
The employer cares that my work performance has not decreased during COVID-19.	3.60	5.00	1.697
The employer cares that my willingness to work does not decrease during COVID-19.	3.58	5.00	1.685
The employer cares that the quality of my work does not decrease during COVID-19.	3.75	5.00	1.696
The employer cares that I do the planned amount of work within a specific time frame during COVID-19.	3.56	5.00	1.724
The employer cares that my focus on achieving successful work results has not decreased during COVID-19.	3.70	5.00	1.733
The employer cares that my accuracy at work has not decreased during COVID-19.	3.71	5.00	1.716
The employer cares that my concentration during work from home has not decreased.	3.53	4.00	1.255

Table 1.4 shows that the means for answers about the employee satisfaction with employer care about their work efficiency indicate that, on average, employees who WFH during the COVID-19 had the highest agreement with: The employer cares that the quality of my work does not decrease during the COVID-19 (mean: 3.75), the employer cares that my accuracy at work has not decreased during the COVID-19 (mean: 3.71), the employer cares that my focus on achieving successful work results has not decreased during the COVID-19 (mean: 3.70), the employer cares that my workability has not decreased during the COVID-19 (mean: 3.68), the employer cares that my work performance has not decreased during the COVID-19 (mean: 3.60), the employer cares that my willingness to work does not decrease during the COVID-19 (mean: 3.58), the employer cares that I do the planned amount of work within a specific time frame during the COVID-19 (mean: 3.56), the employer cares that my concentration during the WFH has not decreased (mean: 3.53). The highest dispersion of employees who WFH during the COVID-19 is observed in the statement: "The employer cares that my focus on achieving successful

work results has not decreased during the COVID-19." The lowest dispersion of employees who WFH during the COVID-19 is observed in the statement: "The employer cares that my concentration during the WFH has not decreased."

TABLE 1.5 Descriptive Statistics for Construct Employee Satisfaction with Employer Care About Their Work Motivation

Statement	Mean	Median	Standard Deviation
The employer gives us compliments for the well-done work.	3.69	4.00	1.175
The employer cares that in the organization prevail respect between leader and employees.	3.98	4.00	0.915
The employer cares that in the organization prevail respect among employees.	3.77	4.00	0.944
The employer cares that in the organization prevail good relationships between leader and employees.	3.76	4.00	1.231
The employer cares that in the organization prevail good relationships among employees.	3.72	4.00	1.244
The employer motivates us with different flexible forms of work during COVID-19.	3.66	4.00	1.239
The employer gives me the possibility of autonomy at work during COVID-19.	3.95	4.00	0.875
The employer allows me to do my work at my own pace during COVID-19.	3.73	4.00	1.205
The employer gives me the possibility for training and education during COVID-19.	3.39	3.00	1.041
The employer allows me to provide diverse tasks.	3.12	3.00	1.323

Table 1.5 shows that the means for answers about employee satisfaction with employer care about their work motivation indicate that, on average, employees who WFH during the COVID-19 had the highest agreement with: The employer cares that in the organization prevail respect between employer and employees (mean: 3.98), the employer gives me the possibility of autonomy at work during the COVID-19 (mean: 3.95), the employer cares that in the organization prevail respect among employees (mean: 3.77), the employer cares that in the organization prevail good relationships between employer and employees (mean: 3.76), the employer allows me to do my work at my own pace during

the COVID-19 (mean: 3.73), the employer cares that in the organization prevail good relationships among employees (mean: 3.72), the employer gives us compliments for the well-done work (mean: 3.69) and the employer motivates us with different flexible forms of work during the COVID-19 (mean: 3.66). On the other hand, results indicate that employees who WFH during the COVID-19 on average had the partial agreement with: the employer gives the possibility for training and education during the COVID-19 (mean: 3.39), the employer allows to provide diverse tasks (mean: 3.12). The highest dispersion of employees who WFH during COVID-19 has been observed in the statement: "The employer allows me to provide diverse tasks." On the other hand, the lowest dispersion of employees who WFH during COVID-19 has been observed in the statement: "The employer gives me the possibility of autonomy at work during COVID-19."

1.4.2 RESULTS OF STRUCTURAL EQUATION MODELING WITH PLS-SEM

1.4.2.1 DIMENSIONALITY, VALIDITY, AND RELIABILITY

In the first step, the measurement model was assessed. Next, psychometric properties (measurement model) of these scales were assessed via evaluation of reliability, convergent validity and discriminant validity of each measurement scale. Fornell and Larcker's (Fornell & Larcker, 1981) assessment criteria were adopted for convergent validity. All item factor loadings should be significant and should exceed 0.70, and the AVE for each construct should exceed 0.50 (Garson, 2016; Henseler et al., 2015). Tables 1.6–1.10 show item factor loadings for each construct (employee satisfaction with working conditions, employee satisfaction with employer care about their well-being, employee satisfaction with employer care about their work engagement, employee satisfaction with employer care about their work efficiency, employee satisfaction with employer care about their work motivation). Table 1.11 shows constructs' reliability and validity.

Table 1.6 shows that item factor loadings for construct employee satisfaction with working conditions were higher than the recommended level of 0.70 and were significant at $p<0.001$. Table 1.7 shows item factor

TABLE 1.6 Item Factor Loadings for the Construct Employee Satisfaction with Working Conditions

Statement	Loadings
I am satisfied with the internal communication.	0.816
I am satisfied with all the information regarding the changes in the organization.	0.852
I am satisfied with enabling flexible working time.	0.910
I am satisfied with the balance between work and private life.	0.734
I am satisfied with the daily schedule of work obligations.	0.782
I am satisfied with the number of different training programs (online).	0.727
I am satisfied with teleworking (work from home).	0.825
I am satisfied with the number of breaks during work which I perform from home.	0.906
I am satisfied with the flexible forms of work that the organization allows me during COVID-19.	0.938
I am satisfied with the leadership.	0.893

loadings for the construct employee satisfaction with employer care about their well-being.

TABLE 1.7 Item Factor Loadings for the Construct Employee Satisfaction with Employer Care About Their Well-Being

Statement	Loadings
When I have work problems, the leader takes time to listen to me.	0.870
The employer is always ready to help me when I have problems.	0.807
The employer cares that I feel good when I perform work from home.	0.772
The employer cares that I am not subject to work stress.	0.809
The employer cares that I do not feel workload.	0.877
The employer cares that I do not feel the behavioral symptoms of burnout.	0.752
The employer cares that I do not feel the physical symptoms of burnout.	0.721
The employer cares that I do not feel the emotional symptoms of burnout.	0.769
The employer organizes different seminars about health, well-being, stress, etc.	0.728

Table 1.7 shows that item factor loadings for construct employee satisfaction with employer care about their well-being were higher than the recommended level of 0.70 and were significant at $p<0.001$. Table 1.8

shows item factor loadings for the construct employee satisfaction with employer care about their work engagement.

TABLE 1.8 Item Factor Loadings for the Construct Employee Satisfaction with Employer Care About Their Work Engagement

Statement	Loadings
The employer cares that I do my work with passion when I perform work from home during COVID-19.	0.794
The employer cares that I am engaged in the quality of my work during COVID-19.	0.757
The employer cares that I am engaged in achieving successful business results during COVID-19.	0.716
The employer cares that I am aware of the importance of innovation for our organization and that I am helping to develop the organization during COVID-19.	0.785
The employer cares that I feel the connection with the organization in which I worked.	0.845
The employer cares that I feel that my work and job are essential.	0.884
The employer takes my business ideas into account.	0.781
The employer cares that I have everything I need to do my job well during COVID-19.	0.792
Over the last year, I had the opportunity to develop.	0.848

Table 1.8 shows that item factor loadings for construct employee satisfaction with employer care about their work engagement were higher than the recommended level of 0.70 and were significant at $p<0.001$. Table 1.9 shows item factor loadings for the construct employee satisfaction with employer care about their work efficiency.

Table 1.9 shows that item factor loadings for construct employee satisfaction with employer care about their work efficiency were higher than the recommended level of 0.70 and were significant at $p<0.001$. Table 1.10 shows item factor loadings for the construct employee satisfaction with employer care about their work motivation.

Table 1.10 shows that item factor loadings for construct employee satisfaction with employer care about their work motivation were higher than the recommended level of 0.70 and were significant at $p<0.001$. Table 1.11 shows the constructs' reliability and validity.

Within the scope of inner consistency with Cronbach's alpha coefficient, the reliability of measurement scales was assessed (Chronbach,

TABLE 1.9 Item Factor Loadings for the Construct Employee Satisfaction with Employer Care About Their Work Efficiency

Statement	Loadings
The employer cares that my workability has not decreased during COVID-19.	0.851
The employer cares that my work performance has not decreased during COVID-19.	0.783
The employer cares that my willingness to work does not decrease during COVID-19.	0.845
The employer cares that the quality of my work does not decrease during COVID-19.	0.877
The employer cares that I do the planned amount of work within a specific time frame during COVID-19.	0.914
The employer cares that my focus on achieving successful work results has not decreased during COVID-19.	0.813
The employer cares that my accuracy at work has not decreased during COVID-19.	0.817
The employer cares that my concentration during work from home has not decreased.	0.779

TABLE 1.10 Item Factor Loadings for the Construct Employee Satisfaction with Employer Care About Their Work Motivation

Statement	Loadings
The employer gives us compliments for the well-done work.	0.723
The employer cares that in the organization prevail respect between leader and employees.	0.756
The employer cares that in the organization prevail respect among employees.	0.734
The employer cares that in the organization prevail good relationships between leader and employees.	0.906
The employer cares that in the organization prevail good relationships among employees.	0.878
The employer motivates us with different flexible forms of work during COVID-19.	0.870
The employer gives me the possibility of autonomy at work during COVID-19.	0.803
The employer allows me to do my work at my own pace during COVID-19.	0.729
The employer gives me the possibility for training and education during COVID-19.	0.712
The employer allows me to provide diverse tasks.	0.716

1951). According to Churchill & Brown (2004), the indicators of highly reliable constructs should be connected, meaning they measure the same latent construct. The alpha coefficient of $\alpha \geq 0.80$ is marked as exemplary; the coefficient in the interval $0.70 \leq \alpha < 0.80$ as very good; the interval $0.60 \leq \alpha < 0.70$ as moderate; and smaller than 0.60 as barely acceptable. Within convergent validity, we examined CR coefficients (>0.7) and AVE (>0.5) and the criterion CR>AVE (Kock, 2016). According to Kock (2016), the AVE for each construct should exceed 0.50.

TABLE 1.11 Constructs' Reliability and Validity

Construct	Cronbach's Alpha	AVE	CR
Employee satisfaction with working conditions	0.941	0.945	0.971
Employee satisfaction with employer care about their well-being	0.925	0.626	0.938
Employee satisfaction with employer care about their work engagement	0.915	0.630	0.931
Employee satisfaction with employer care about their work efficiency	0.923	0.720	0.939
Employee satisfaction with employer care about their work motivation	0.854	0.629	0.891

Note: CR: Composite reliability; AVE: Average variance extracted.

Table 1.11 shows that Cronbach's alphas indicate that the scales were reliable for all measurement scales. In addition, composite reliabilities (CR) for all five constructs are greater than 0.7, and the values of AVE for all four constructs are greater than 0.5. Thus, we confirmed the convergent validity for all the constructs studied because all CR values were higher than AVE values.

Discriminant validity between constructs was assessed following Fornell and Larcker's (Fornell & Larcker, 1981) recommendation that the square root of AVE for each construct should exceed the bivariate correlations between this construct and all other constructs. The inter-construct correlation matrix in Table 1.12 shows that the principal diagonal elements (square root AVE) exceed non-diagonals elements in the same row or columns (bivariate correlations), demonstrating that the discriminate validity of all scales is also adequate. Therefore, Fornell and Larcker's (Fornell & Larcker, 1981) criterion was used to test for discriminant validity, which means that the square roots of AVE should be higher than the latent variable correlations.

TABLE 1.12 Square Roots of the AVE and Correlations Among the Latent Variables

Constructs	Employee Satisfaction with Working Conditions	Employee Satisfaction with Employer care About Their Work Efficiency	Employee Satisfaction with Employer Care About Their Work Engagement	Employee Satisfaction with Employer Care About Their Work Motivation	Employee Satisfaction with Employer Care About Their Well-Being
Employee satisfaction with working conditions.	0.972*	–	–	–	–
Employee satisfaction with employer care about their work efficiency.	0.455	0.849*	–	–	–
Employee satisfaction with employer care about their work engagement.	0.672	0.652	0.794*	–	–
Employee satisfaction with employer care about their work motivation.	0.465	0.341	0.673	0.793*	–
Employee satisfaction with employer care about their well-being.	0.724	0.595	0.700	0.361	0.791*

Note: * Square roots of AVE.

Table 1.12 indicates that Fornell and Larcker's (Fornell & Larcker, 1981) criterion was achieved in all cases since all square roots of AVE were higher than the latent variable correlations.

1.4.2.2 PLS-SEM RESULTS

Structural path coefficients were analyzed using the significance level at 0.05. The R square and Adjusted R square were used to measure the total magnitude of the effect size for the structural equation model, with the results above the cut-offs of 0.67, 0.33, and 0.19 being substantial, moderate, and weak, respectively (Chin, 1998). Table 1.13 shows the goodness of fit for the structural model.

TABLE 1.13 The Goodness of Fit for the Structural Model Through

Constructs	R Square	Adjusted R Square
Employee satisfaction with working conditions.	–	–
Employee satisfaction with employer care about their work motivation.	–	–
Employee satisfaction with employer care about their work efficiency.	0.484	0.482
Employee satisfaction with employer care about their work engagement.	0.694	0.693
Employee satisfaction with employer care about their well-being.	0.654	0.653

The results in Table 1.13 indicates that R square values for all three endogenous variables (employee satisfaction with employer care about their work efficiency, employee satisfaction with employer care about their work engagement, employee satisfaction with employer care about their well-being) were larger than 0.4. Further, Table 1.14 shows the fit indices for PLS-SEM where the NFI and the SRMR index (SRMR index) were used. The model adaptation was assessed by the SRMR. Henseler et al. (2014) introduce the SRMR as a goodness of fit measure for PLS-SEM that can be used to avoid model misspecification. According to Garson (2016), the model fits well when SRMR is lower than 0.10. According to Hu & Bentler (1999), a value less than 0.10 or 0.08 are considered a good fit (Hu & Bentler, 1999). Also, a model is regarded as acceptable if the NFI exceeds 0.90 (Byrne, 1994). Table 1.14 shows the fit indices for PLS-SEM.

TABLE 1.14 Fit Indices for PLS-SEM

Index	Value
NFI	0.936
SRMR	0.078

Note: NFI: normed fit index; SRMR: standardized root mean square residual index.

Results in Table 1.14 shows that NFI and SRMR criterion was achieved. Thus, the NFI index exceeds 0.90 (NFI = 0.936) and also, SRMR index is less than 0.08 (SRMR = 0.078).

Statistically significant relationships among the constructs were tested for hypothesis H1, H2, H3, H4, H5, and H6. Table 1.15 and Figure 1.2 show the statistically significant relationships among the constructs.

TABLE 1.15 Standardized Regression Paths

Hypothesis		Standardized Regression Paths (β)
H1	Employee satisfaction with working conditions → Employee satisfaction with employer care about their work efficiency.	0.794 ($p<0.01$)
H2	Employee satisfaction with working conditions → Employee satisfaction with employer care about their well-being.	0.809 ($p<0.01$)
H3	Employee satisfaction with employer care about their well-being → Employee satisfaction with employer cares about their work engagement.	0.710 ($p<0.01$)
H4	Employee satisfaction with employer care about their work engagement → Employee satisfaction with employer care about their work efficiency.	0.816 ($p<0.01$)
H5	Employee satisfaction with employer care about their work motivation → Employee satisfaction with employer cares about their work engagement.	0.723 ($p<0.01$)
H6	Employee satisfaction with employer care about their work motivation → Employee satisfaction with employer care about their work efficiency.	0.731 ($p<0.01$)

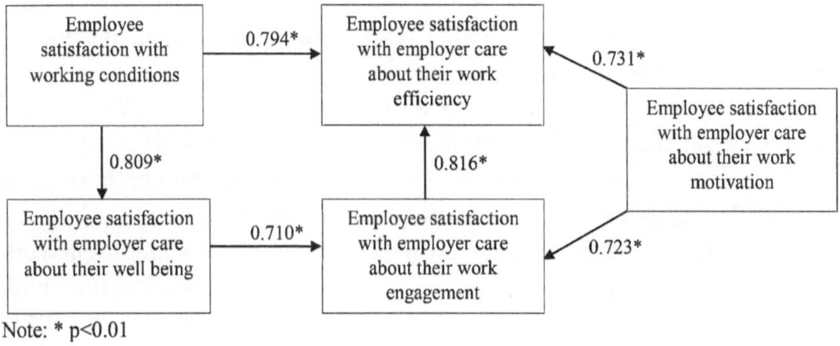

FIGURE 1.2 Results of structural model analysis for employee satisfaction with leadership during the COVID-19 pandemic.

Note: * $p<0.01$.

As can be observed from Table 1.15 and Figure 1.2, the results showed that employee satisfaction with working conditions positively affects employee satisfaction with employer care about their work efficiency among employees who WFH during COVID-19 ($\beta = 0.794$, $p<0.01$). Additionally, employee satisfaction with working conditions positively affects employee satisfaction with employer care about their well-being among employees who WFH during COVID-19 ($\beta = 0.809$, $p<0.01$). The results showed that employee satisfaction with employer cares about their well-being has a positive effect on employee satisfaction with employer care about their work engagement among employees who WFH during the COVID-19 ($\beta = 0.710$, $p<0.01$). Also, employee satisfaction with employer care about their work engagement has a positive effect on employee satisfaction with employer care about their work efficiency among employees who WFH during COVID-19 ($\beta = 0.816$, $p<0.01$). The results show that employee satisfaction with employer cares about their work motivation has a positive effect on employee satisfaction with employer care about their work engagement among employees who WFH during the COVID-19 ($\beta = 0.732$, $p<0.01$). In addition, employee satisfaction with employer care about their work motivation positively affects employee satisfaction with employer care about their work efficiency among employees who WFH during the COVID-19 ($\beta = 0.731$, $p<0.01$). Therefore, we confirmed hypotheses H1 (H1: Employee satisfaction with working conditions has a positive direct effect on employee satisfaction with employer care about their work efficiency, among employees who WFH during the COVID-19), H2 (H2: Employee satisfaction with working conditions has a positive direct effect on employee satisfaction with employer care about their well-being, among employees who WFH during the COVID-19), H3 (H3: Employee satisfaction with employer care about their well-being has a positive direct effect on employee satisfaction with employer care about their work engagement, among employees who WFH during the COVID-19), H4 (H4: Employee satisfaction with employer care about their work engagement has a positive direct effect on employee satisfaction with employer care about their work efficiency, among employees who WFH during the COVID-19), H5 (H5: Employee satisfaction with employer care about their work motivation has a positive direct effect on employee satisfaction with employer care about their work engagement, among employees who WFH during the COVID-19) and H6 (H6: Employee satisfaction with employer care about their work motivation has a positive direct effect on employee satisfaction with employer care about their work efficiency, among employees who WFH during the COVID-19).

1.5 DISCUSSION

Nowadays, healthy organizations are essential because they promote their employees' physical, emotional, mental, and spiritual aspects (Abbas & Zhiqiang, 2020; Caringal-Go et al., 2021; supplemented). The organizations that are aware of the importance of health and good well-being of their employees balance economic value-added, customer value-added and people value-added. In those organizations, profits are not something to be achieved at the expense of employees. The hidden value in an organization is usually thought of as soft rather than hard assets. Organization' value is increasingly being driven by employees and their ideas, as opposed to hard assets. Thus, employees not only create value, but they are also highly valued themselves (Budhwar et al., 2002; Zimmerman, 2001; Adikaram et al., 2021). This is especially important during COVID-19 and after COVID-19 because COVID-19 requires new work strategies and approaches to which employees must adapt (Anderson & Kelliher, 2020). During the COVID-19 pandemic, one of the most challenging issues confronting leaders is how to improve employee well-being, work efficiency, work engagement and work motivation among their employees during the COVID-19 (Bartik et al., 2020; Bhattarai, 2020; Caringal-Go et al., 2021).

Based on results, we found that *employee satisfaction with working conditions* has a positive direct effect: (i) on employee satisfaction with employer care about their work efficiency; and (ii) employee satisfaction with employer care about their well-being among employees who WFH during the COVID-19. Our findings are in line with Liang et al. (2020); Haque (2021); Bhattarai (2020); Diab-Bahman & Al-Enzi (2020); and Nagel (2020). Those authors found out that employee satisfaction with working conditions during COVID-19 positively impacted employee satisfaction with employers caring about their work efficiency and employee satisfaction with employers caring about their well-being. According to Nagel (2020); Bhattarai (2020); and Diab-Bahman & Al-Enzi (2020), a new working condition during COVID-19 as teleworking increases the autonomy and flexible working time among employees who WFH during COVID-19, and this may lead to a higher level of satisfaction with employer care about their work efficiency and well-being. Furthermore, Haque (2021) emphasizes that employee well-being has been recognized as a significant motivating factor for higher organizational performance (Li et al., 2011). In addition, our research results show that employee satisfaction with working conditions is reflected mostly; (iii) in satisfaction with the flexible forms of work during the COVID-19; (iv) with enabling flexible working time; (v) with the number of breaks during work; (vi) with the leadership; (vii) with all the information regarding the changes in the organization; and (viii) with the internal communication. Also, employees who WFH during the COVID-19 are satisfied; (ix) with the daily schedule of work obligations; (x) with teleworking (WFH); (xi) with the balance between work and private life; and (xii) with the number of different training programs (online). However, their value of the agreement with the latter is lower.

To positively affect employee work efficiency and well-being, we *recommend* that an employer supports them to maintain their work-life balance during the COVID-19. Namely, during COVID-19, many employees are facing increased care for the elderly and children. Also, much emphasis should be given to internal communication between the employer and employee, because based on this, an employer can identify which employees are in distress. The work tasks should be adapted to those employees who are in distress and unable to reconcile work and private. Therefore, employers need to continue pushing for more transparency of employees and communicate with their employees. Employers need to be ready to listen to what challenges or problems employees face and find an appropriate solution. This will improve the employee's work efficiency as well as their well-being. In addition, employees have to adapt to new ways of working and new digital tools, which causes additional stress. Providing the right digital tools that work and integrate seamlessly with other programs and devices is another way that organization can contribute to good well-being and better work efficiency.

Based on results, we found that *employee satisfaction with employer care about their well-being* has a positive direct effect on employee satisfaction with employer care about their work engagement among employees who WFH during the COVID-19. Our findings are in line with Adikaram et al. (2021); Mani & Mishra (2020); and Douglas & Roberts (2020), in which authors emphasize that organization in which employers care for their employees in terms of their health and well-being this lead to higher work engagement among employees during the COVID-19. When employees feel good and are engaged, this leads to great organizational competitiveness during COVID-19 because engaged employees are the key to organizational success. In addition, the results show that employee satisfaction with employer care about their well-being is most reflected in satisfaction with five statements: "When I have work problems, the employer takes time to listens to me," "The employer is always ready to help me when I have problems," "The employer cares that I feel good when I perform WFH," "The employer cares that I am not subject to work stress" and "The employer cares that I do not feel workload." On the other hand, results indicate that employees who WFH during the COVID-19 on average had the partial agreement with four statements: "The employer cares that I do not feel the behavioral symptoms of burnout," "The employer cares that I do not feel the physical symptoms of burnout," "The employer cares that I do not feel the emotional symptoms of burnout," and "The employer organizes different seminars about health, well-being, stress, etc."

Also, the results indicate that *employee satisfaction with employer care about their work engagement* is most reflected in: (i) satisfaction with employer care that employees have everything to do their job well during the COVID-19; (ii) with employer care that employees are engaged to the quality of their work during the COVID-19; (iii) the satisfaction that the employer takes employees business ideas into account; (iv) satisfaction with employer care that employees are engaged to achieve successful business results during the COVID-19; (v) with employer care that employees feel that their work and job are essential; (vi) with employer care that employees feel the connection with the organization; (vii) with employer care that employees do their work with passion when they perform WFH during the COVID-19; and (viii) with employer cares that employees are aware of the importance of innovation for organization and that they help to develop the organization during the COVID-19. But, on the other hand, employees partly agree with the opportunity to develop over the last year.

The employees are an essential part of the organization. Therefore, for better well-being and higher work engagement during COVID-19, we *recommend*: strategies to prevent stress and burnout among employees who WFH during COVID-19 need to be applied. Also, employers need to be educated about stress and burnout and their causes and symptoms to be able to help their employees as effectively as possible. Additionally, employers should be sensitive to employees' needs and pay attention to the factors that contribute to the greater engagement of each employee that works from home during COVID-19. Each employee has to be considered individually because the level of work engagement of each employee is different. Finally, employers need to invest time in open conversations about what has been lost in the COVID-19 pandemic. While conversations about the emotional toll of the COVID-19 pandemic may seem unnecessary, they help strengthen ties with employees who appreciate leaders' openness.

Based on the results, we found that *employee satisfaction with employer care about their work engagement* has a positive direct effect on employee satisfaction with employer care about their work efficiency among employees who WFH during COVID-19. Our findings are in line with Zhang et al. (2020); Li et al. (2020); and Haque (2021). Authors found that appropriate and suitable leadership during COVID-19 positively impacted work efficiency among employees who WFH during COVID-19. Also, employee satisfaction with employer care about their work efficiency is higher if employer cares about their work engagement. The results of our research indicate that employee satisfaction with employer care about their work efficiency is most reflected in satisfaction: (i) with the employer care that the quality of employees work does not decrease during the COVID-19; (ii) with the employer care that employees accuracy at work has not decreased during the COVID-19; (iii) with employer care that employees focus on achieving successful work results has not decreased during the COVID-19; (iv) with employer care that employees workability has not decreased during the COVID-19; (v) with employer care that employees work performance has not decreased during the COVID-19; (vi) with employer care that employees willingness to work does not decrease during the COVID-19; (vii) with employer care that employees do the planned amount of work within a specific time frame during the COVID-19; and (viii) with employer care that employees concentration during the WFH has not decreased. The research of Morikawa (2020) indicates that the work efficiency of Japanese employees who WFH during the COVID-19 pandemic is, on average, 30–40% lower than that at the workplace. Also, Bartik et al. (2020), using data from a survey of small and medium-sized organizations in the US, indicate decreased work efficiency of about 20% on average.

Work efficiency is crucial for the successful results and the competitiveness of the organization. Therefore, we *recommend* that the employer pay more attention to the concentration of employees who WFH during COVID-19 and motivate them accordingly. Also, due to increased stress among employees who WFH during COVID-19, the employer should care about their willingness to work and do the planned work within a specific time frame. If some employees who WFH are overworked, the employer should allocate their work responsibilities. An employer should know employees' skills and behavioral styles and, on that basis, use appropriate communication with each employee. Also, the business goal or work obligations should be clearly defined and achievable in this epidemiological situation. The employer should ensure that employees' assignments are as straightforward as possible and let them know their expectations.

Based on results, we found that *employee satisfaction with employer care about their work motivation* has a positive direct effect on employee satisfaction with employer care about their work engagement and employee satisfaction with employer care about their work efficiency among employees who WFH during the COVID-19. Our findings are in line with Tovmasyan & Minasyan (2020); Wiradadi et al. (2020); Toscano & Zappalà (2020); Dwivedi et al. (2020) in which authors found out that employees who feel supported by the organization are more motivated to do their best work which leads to higher work engagement and work efficiency. Moreover, COVID-19 has changed businesses operate, and the way employees work. Organizations and employees have recently seen dramatic changes in working patterns, reflected in employee motivation, engagement, and efficiency. Now, more than ever, employer cares about employee motivation is key to organizational success and competitiveness. The results of our research show that employee satisfaction with employer cares about their work motivation is reflected in motivation with: (i) the employer care that in the organization prevail respect between employer and employees, the possibility of autonomy at work during the COVID-19; (ii) the employer care that in the organization prevail respect among employees; (iii) the employer care that in the organization prevail good relationships between employer and employees, the possibility to do work at own pace during the COVID-19; (iv) the employer care that in the organization prevail good relationships among employees, the compliments for the well-done work, and the possibility of different flexible forms of work during the COVID-19. On the other hand, results show that employees who WFH during COVID-19 had partial motivation to train and educate during COVID-19 and the opportunity to provide diverse tasks.

Therefore, we *recommend*: The employer should: (i) encourage career development and training (online); and (ii) consider reduced hours for a working parent, which leads to higher employee engagement and efficiency. Also, the employer should recognize their employees for work well done, which will make employees feel appreciated and encourage them to continue increasing their productivity. Furthermore, the employer should recognize employees hard work and gives them compliments or reward employees. The organization should create a sense of belonging. The more information the employer share with employees, the more accessible employees can cope with work during COVID-19. Also, the employer should ask employees for business suggestions because this reinforces employee engagement, efficiency, and satisfaction with the employer care about them. In addition, the employer should show employees that their work is meaningful by sharing excellent outcomes. According to the Global Talent Survey (2021), the COVID-19 pandemic has radically changed the fundamental values and preferences in the workplaces of Slovenian employees. Slovenian employees motivate most the excellent relationship with co-workers, the balance between work and private life, a good relationship with employer and learning and developing skills. The salary is not one of their priorities; the employer's financial stability is more important to them. Also, the research showed that globally, after the end of the pandemic, 89% of participants expect the possibility of combining flexible and permanent work, while for Slovenes, this percentage is lower, namely 72%. Such a working condition is advocated mainly by employees in digitization and automation, staffing, administration, secretariat, and information technology. It is now up to organizations to accept what they have learned and adapt long-term solutions that align with expectations after the pandemic. At the same time, employers will need to ensure that employees can maintain work-life balance even with changed ways of working. The confirmed hypotheses are clearly shown in Table 1.16.

TABLE 1.16 Assessed and Confirmed Hypotheses

Hypotheses	Confirmation
➢ **H1:** *Employee satisfaction with working conditions has a positive direct effect on employee satisfaction with employer care about their work efficiency, among employees who work from home during COVID-19.*	✓
➢ **H2:** *Employee satisfaction with working conditions has a positive direct effect on employee satisfaction with employer care about their well-being, among employees who work from home during COVID-19.*	✓
➢ **H3:** *Employee satisfaction with employer care about their well-being has a positive direct effect on employee satisfaction with employer care about their work engagement, among employees who work from home during COVID-19.*	✓
➢ **H4:** *Employee satisfaction with employer care about their work engagement has a positive direct effect on employee satisfaction with employer care about their work efficiency, among employees who work from home during COVID-19.*	✓
➢ **H5:** *Employee satisfaction with employer care about their work motivation has a positive direct effect on employee satisfaction with employer care about their work engagement, among employees who work from home during COVID-19.*	✓
➢ **H6:** *Employee satisfaction with employer care about their work motivation has a positive direct effect on employee satisfaction with employer care about their work efficiency, among employees who work from home during COVID-19.*	✓

1.6 CONCLUSION

The COVID-19 pandemic has a high impact on the sustainability of organizations at all business levels. At this time, organizations have to fight more for their competitiveness. Employee satisfaction with appropriate leadership during COVID-19 plays a vital role in achieving successful business results and competitive advantage. According to Li et al. (2020); Zhang et al. (2020); and Coun et al. (2019), leadership is an important organizational resource, which leads to higher work efficiency, work engagement, work motivation and also, well-being among employees. This is especially important during COVID-19 among employees who WFH. Our findings shed light on the need for organizations to be aware of the extent and factors associated with appropriate working conditions, employee well-being, employee engagement, employee efficiency and employee motivation. Also, our research highlights the effects in employee satisfaction with leadership, which is influencing an organization's future competitiveness, through five constructs: (i) employee satisfaction with working conditions; (ii) employee satisfaction with employer care about their well-being; (iii) employee satisfaction with employer care about their work engagement; (iv) employee satisfaction with employer care about their work efficiency; and (v) employee satisfaction with employer care about their work motivation during the COVID-19 pandemic.

Additionally, the chapter highlights the developed multidimensional model and the strength of the relationships among constructs in the model. Based on the results, we concluded that employee satisfaction with working conditions has a positive direct effect on employee satisfaction with employer care about their work efficiency and employee satisfaction with employer care about their well-being among employees who WFH during the COVID-19. Also, our research shows that employee satisfaction with employer care about their well-being has a positive direct effect on employee satisfaction with employer care about their work engagement. Further, employee satisfaction with employer care about their work engagement has a positive direct effect on employee satisfaction with employer care about their work efficiency, among employees who WFH during the COVID-19. Furthermore, based on the results, we concluded that employee satisfaction with employer care about their work motivation: (i) has a positive direct effect on employee satisfaction with employer care about their work engagement; as well as (ii) on employee satisfaction with employer care about their work efficiency, among employees who WFH during the COVID-19. The limitation of the research is reflected in organizations in Slovenia. In our research, we limited on five constructs: (i) employee satisfaction with working conditions; (ii) employee satisfaction with employer care about their well-being; (iii) employee satisfaction with employer care about their work engagement; (iv) employee satisfaction with employer care about their work efficiency; and (v) employee satisfaction with employer care about their work motivation during the COVID-19 pandemic. We are researchers from two different countries, Slovenia and Croatia; therefore, we propose analyzing these constructs in organizations in Croatia for further research. Further research possibilities we also see in examining the differences in these five constructs among Croatian and Slovenian employees.

KEYWORDS

- **average variance extracted**
- **competitiveness**
- **composite reliabilities**
- **COVID-19**
- **employees**
- **leadership**
- **mental health**
- **quality of life**
- **work-life balance**

REFERENCES

Abbas, S. M., & Zhiqiang, L., (2020). COVID-19, mental well-being and work engagement: The psychological resilience of senescent workforce. *International Journal of Research in Business & Social Science, 9*(4), 356–365.

Adikaram, A. S., Naotunna, N. P. G. S. S., & Priyankara, H. P. R., (2021). Battling COVID-19 with human resource management bundling. *Employee Relations*.

Amoah, C., & Simpeh, F., (2020). Implementation challenges of COVID-19 safety measures at construction sites in South Africa. *Journal of Facilities Management, 19*(1), 111–128.

Anderson, D., & Kelliher, C., (2020). Enforced remote working and the work-life interface during lockdown. *Gender in Management, 35*(7, 8), 677–683.

Baert, S., Lippens, L., Moens, E., Sterkens, P., & Weytjens, J., (2020). *The COVID-19 Crisis and Telework: A Research Survey on Experiences, Expectations and Hopes, (Working Paper)*. Available at: http://wps-feb.ugent.be/Papers/wp_20_996.pdf (accessed on 7 February 2023).

Bartik, A. W., Cullen, Z. B., Glaeser, E. L., Luca, M., & Stanton, C. T., (2020). *What Jobs are Being Done at Home During the COVID-19 Crisis? Evidence from Firm-Level Surveys*. NBER Working Paper 27422. Available at: https://www.nber.org/system/files/working_papers/w27422/w27422.pdf (accessed on 7 February 2023).

Bhattarai, M., (2020). *Working from Home and Job Satisfaction During the Pandemic Times*. Canada: Independent Publication.

Budhwar, P., Crane, A., Davies, A., Delbridge, R., Edwards, T., Ezzamel, M., Harris, L., et al., (2002). Organizing/theorizing: Developments in organization theory and practice. *Management Research News, 25*(8–10), 1–193.

Byrne, B. M., (1994). *Structural Equation Modeling with EQS and EQS/Windows*. Thousand Oaks, CA: Sage Publications.

Caringal-Go, J. F., Teng-Calleja, M., Franco, E. P., Manaois, J. O., & Zantua, R. M. S., (2021). Crisis leadership from the perspective of employees during the COVID-19 pandemic. *Leadership & Organization Development Journal*.

Chanana, N., & Sangeeta, M., (2020). *Employee Engagement Practices During COVID-19 Lockdown*. Available at: https://onlinelibrary.wiley.com/doi/epdf/10.1002/pa.2508 (accessed on 7 February 2023).

Chin, W. W., (1988). Issues and opinion on structural equation modeling. *MIS Quarterly, 22*(1), 1–10.

Chronbach, L. J., (1951). Coefficient alpha and the internal structure of tests. *Psychometrika, 16*, 297–334.

Chung, H. J., (2020). *Return of the 1950s Housewife? How to Stop Coronavirus Lockdown Reinforcing Sexist Gender Roles*. Available at: https://theconversation.com/return-of-the-1950s-housewifehow-to-stop-coronavirus-lockdown-reinforcing-sexist-gender-roles-134851 (accessed on 7 February 2023).

Churchill, G. A., & Brown, T. J., (2004). *Basic Marketing Research*. Thomson: Mason, OH, USA.

Conrad, D., Ghosh, A., & Isaacson, M., (2015). Employee motivation factors: A comparative study of the perceptions between physicians and physician leaders. *International Journal of Public Leadership, 11*(2), 92–106.

Coun, M., Peters, P., & Blomme, R., (2019). Let's share! The mediating role of employees' self-determination in the relationship between transformational and shared leadership and perceived knowledge sharing among peers. *European Management Journal, 37*(4), 481–491.

Dalkrani, M., & Dimitriadis, E., (2018). The effect of job satisfaction on employee commitment. *International Journal of Business and Economic Sciences Applied Research, 11*(3), 16–23.

Davidescu, A. A., Apostu, S. A., Paul, A., & Casuneanu, I., (2020). Work flexibility, job satisfaction, and job performance among Romanian employees – implications for sustainable human resource management. *Sustainability, 12*(15), 1–53.

Diab-Bahman, R., & Al-Enzi, A., (2020). The impact of COVID-19 pandemic on conventional work settings. *International Journal of Sociology and Social Policy, 40*(9, 10), 909–927.

Díaz-Carrión b, R., Navajas-Romeroa, V., & Casas-Rosalc, J. C., (2020). Comparing working conditions and job satisfaction in hospitality workers across Europe. *International Journal of Hospitality Management, 90*, 1–13.

Douglas, S., & Roberts, R., (2020). Employee age and the impact on work engagement. *Strategic HR Review, 19*(5), 209–213.

Dwivedi, P., Chaturvedi, V., & Vashist, J. K., (2020). Transformational leadership and employee efficiency: Knowledge sharing as mediator. *Benchmarking: An International Journal, 27*(4), 1571–1590.

Fornell, C., & Larcker, D. F., (1981). Evaluating structural equation models with unobservable variables and measurement error. *Journal of Marketing Research, 18*, 39–50.

Garson, G. D., (2016). *Partial Least Squares: Regression & Structural Equation Models*. Garson and Statistical Associates Publishing: Asheboro, NC, USA.

Gavidia, M., (2020). *How Has COVID-19 Affected Mental Health, Severity of Stress Among Employees?* Available at: https://www.ajmc.com/newsroom/how-has-covid19-affected-mental-health-severity-of-stress-among-employees (accessed on 7 February 2023).

Global Talent Survey, (2021). *2021 Global Talent Trends Study.* Available at: https://www.mercer.com/our-thinking/career/global-talent-hr-trends.html (accessed on 7 February 2023).

Hair, J. F., Hult, G. T. M., Ringle, C. M., & Sarstedt, M., (2016). *A Primer on Partial Least Squares Structural Equation Modeling (PLS-SEM)*. Sage: Thousand Oaks, CA, USA.

Hamel, L., & Salganicoff, A., (2020). *Is There a Widening Gender Gap in Coronavirus Stress?* Washington, DC: Kaiser Family Foundation. Available at: https://www.kff.org/policy-watch/is-there-widening-gender-gap-in-coronavirus-stress/ (accessed on 22 February 2023).

Haque, A., (2021). The COVID-19 pandemic and the role of responsible leadership in health care: Thinking beyond employee well-being and organizational sustainability. *Leadership in Health Services, 34*(1), 52–68.

He, J., Mao, Y., Morrison, A. M., & Coca-Stefaniak, J. A., (2020). On being warm and friendly: The effect of socially responsible human resource management on employee fears of the threats of COVID-19. *International Journal of Contemporary Hospitality Management, 33*(1), 346–366.

Henseler, J., Dijkstra, T. K., Sarstedt, M., Ringle, C. M., Diamantopoulos, A., Straub, D. W., Ketchen, D. J., et al., (2014). Common beliefs and reality about partial least squares: Comments on Rönkkö & Evermann. *Organizational Research Methods, 17*(2), 182–209.

Henseler, J., Ringle, C. M., & Sarstedt, M., (2015). A new criterion for assessing discriminant validity in variance-based structural equation modelling. *Journal of the Academy of Marketing Science, 43*(1), 115–135.

Hu, L. T., & Bentler, P. M., (1998). Fit indices in covariance structure modeling: Sensitivity to under parameterized model misspecification. *Psychological Methods, 3*(4), 424–453.

Jaiswal, A., & Dyaram, L., (2019). Perceived diversity and employee well-being: Mediating role of inclusion. *Personnel Review, 49*(5), 1121–1139.

Joo, B. K., & Lee, I., (2017). Workplace happiness: Work engagement, career satisfaction, and subjective well-being. *Evidence-based HRM, 5*(2), 206–221.

Kapoor, V., Yadav, J., Bajpai, L., & Srivastava, S., (2021). Perceived stress and psychological well-being of working mothers during COVID-19: A mediated moderated roles of teleworking and resilience. *Employee Relations*.

Keeman, A., Naswall, K., Malinen, S., & Kuntz, J., (2017). Employee well-being: Evaluating a well-being intervention in two settings. *Frontiers in Psychology, 8*, 1–14.

Kock, N., (2016). Advantages of nonlinear over segmentation analyses in path models. *International Journal of e-Collaboration, 12*(4), 1–6.

Kumar, A., Luthra, S., Mangla, S. K., & Kazançoglu, Y., (2020). COVID-19 impact on sustainable production and operations management. *Sustainable Operations and Computers, 1*, 1–7.

Lapowsky, I., (2020). For working parents in the pandemic, a survey finds dads have it easier than moms. Available at: https://www.protocol.com/parents-working-from-home-survey (accessed on 7 February 2023).

Lee, H., (2021). Changes in workplace practices during the COVID-19 pandemic: The roles of emotion, psychological safety and organization support. *Journal of Organizational Effectiveness: People and Performance, 8*(1), 97–128.

Li, W., Bhutto, T. A., Wang, X., Maitlo, Q., Zafar, A. U., & Bhutto, N. A., (2020). Unlocking employees' green creativity: The effects of green transformational leadership, green intrinsic, and extrinsic motivation. *Journal of Cleaner Production, 255*(10), 1–10.

Liang, M., Xin, Z., Yan, D. X., & Jianxiang, F., (2020). How to improve employee satisfaction and efficiency through different enterprise social media use. *Journal of Enterprise Information Management, 34*(3), 922–947.

Madero, G. S., Ortiz, M. O. E., Ramírez, J., & Olivas-Luján, M. R., (2020). Stress and myths related to the COVID-19 pandemic's effects on remote work. *Management Research, 18*(4), 401–420.

Magnavita, N., Tripepi, G., & Di Prinzio, R. R., (2020). Symptoms in health care workers during the COVID-19 epidemic. A cross-sectional survey. *International Journal of Environmental Research and Public Health, 17*(14), 1–15.

Mani, S., & Mishra, M., (2020). Non-monetary levers to enhance employee engagement in organizations – "GREAT" model of motivation during the COVID-19 crisis. *Strategic HR Review, 19*(4), 171–175.

Mensah, E. B. K., & Tawiah, K. A., (2016). Employee motivation and work performance: A comparative study of mining companies in Ghana. *Journal of Industrial Engineering and Management, 9*(2), 255–309.

Meyer, B., Zill, A., Dilba, D., Gerlach, R., & Schumann, S., (2021). Employee psychological well-being during the COVID-19 pandemic in Germany: A longitudinal study of demands, resources, and exhaustion. Available at: https://onlinelibrary.wiley.com/doi/10.1002/ijop.12743 (accessed on 7 February 2023).

Mgammal, M. H., & Al-Matari, E. M., (2021). Survey data of coronavirus (COVID-19) thought concern, employees' work performance, employees background, feeling about job, work motivation, job satisfaction, psychological state of mind and family commitment in two middle east countries. *Data in Brief, 34*, 1–5.

Millard, E., & Blackburn, S., (2020). *COVID Stress is Stifling Employee Engagement, New Study Suggests*. Available at: https://www.verywellmind.com/covid-stress-is-stifling-employee-engagement-5084937 (accessed on 7 February 2023).

Morikawa, M., (2020). Productivity of working from home during the COVID-19 pandemic: Evidence from an employee survey. *Covid Economics, 49*, 132–147.

Nagel, L., (2020). The influence of the COVID-19 pandemic on the digital transformation of work. *International Journal of Sociology and Social Policy, 40*(9, 10), 861–875.

Nguyen, A., (2020). On the clock and at home: Post-COVID-19 employee monitoring in the workplace. *People and Strategy, 43*, 1–3.

O'Moore, M., & Lynch, J., (2007). Leadership, working environment and workplace bullying. *International Journal of Organization Theory & Behavior, 10*(1), 95–117.

Robertson, I. T., & Cooper, C. L., (2010). Full engagement: The integration of employee engagement and psychological well-being. *Leadership & Organization Development Journal, 31*(4), 324–336.

Rudolph, C. W., Allan, B., Clarke, M., Hertel, G., Hirschi, A., Kunze, F., Shockley, K., et al., (2020). Pandemics: Implications for research and practice in industrial and organizational psychology. *Industrial and Organizational Psychology: Perspectives on*

Science and Practice, Forthcoming. Available at: https://psyarxiv.com/k8us2 (accessed on 7 February 2023).

Shen, H., Fu, M., Pan, H., Yu, Z., & Chen, Y., (2020). The impact of the COVID-19 pandemic on firm performance. *Emerging Markets Finance and Trade, 56*(10), 2213–2230.

Toscano, F., & Zappalà, S., (2020). Social isolation and stress as predictors of productivity perception and remote work satisfaction during the COVID-19 pandemic: The role of concern about the virus in a moderated double mediation. *Sustainability, 12*(23), 1–14.

Tovmasyan, G., & Minasyan, D., (2020). The impact of motivation on work efficiency for both employers and employees also during COVID-19 pandemic: Case study from Armenia. *Business Ethics and Leadership, 4*(3), 25–35.

Tuzovic, S., & Kabadayi, S., (2020). The influence of social distancing on employee well-being: A conceptual framework and research agenda. *Journal of Service Management, 32*(2), 145–160.

U.S. Work from Home Survey, (2020). Gensler Research Institute. Available at: https://www.gensler.com/us-wfh-survey-2020-wfh-reinforces-the-benefits-of-the-office (accessed on 7 February 2023).

Uddin, M., (2020). *Addressing Work-Life Balance Challenges of Working Women During COVID-19 in Bangladesh*. Available at: https://onlinelibrary.wiley.com/doi/epdf/10.1111/issj.12267 (accessed on 7 February 2023).

Wiradendi, W. C., Solikhah, D., Susita, D., & Martono, S., (2020). How to maintain employee motivation amid the COVID-19 virus pandemic. *International Journal of Economics and Business Administration, 8*(4), 78–86.

Wong, A. H. K., Cheung, J. O., & Chen, Z., (2020). Promoting effectiveness of "working from home": Findings from Hong Kong working population under COVID-19. *Asian Education and Development Studies, 10*(2), 210–228.

Zhang, J., Xie, C., Wang, J., Morrison, A. M., & Coca-Stefaniak, J. A., (2020). Responding to a major global crisis: The effects of hotel safety leadership on employee safety behavior during COVID-19. *International Journal of Contemporary Hospitality Management, 32*(11), 3365–3389.

Zimmerman, E., (2001). What are employees worth? *Workforce, 80*(2), 1–32.

Žnidaršič, J., & Bernik, M., (2021). Impact of work-family balance results on employee work engagement within the organization: The case of Slovenia. *PLoS One, 16*(1), 1–17.

CHAPTER 2

Hybrid Working: The Future of Organizations

TARUN KUMAR SHARMA

Assistant Professor, Department of Psychology, Mohanlal Sukhadia University, Udaipur, Rajasthan, India

ABSTRACT

COVID-19 pandemic is considered a massive unplanned experiment in the world of work. The organizations were able to sustain its functioning by adopting work-from-home (WFH) policy. The experiences of remote working led us to understand that work can be performed with optimum productivity in WFH mode as well. The organizations like Microsoft and others have done extensive surveys to understand hybrid working perceptions among the employees. Hybrid working, its models and how different occupations/organizations differ in the possibility of adopting hybrid work arrangements is discussed.

Surveys such as Work Trend Index 2021 by Microsoft have revealed that the future of organizations is Hybrid work. The other surveys Riverbed/Aternity Hybrid Work Global Survey 2021, KPMG CEO 2021 outlook survey, NASSCOM have also reported the similar findings. The perspectives of Satya Nadella, Sunder Pichai, Mark Zuckerberg, Mark Dixon, and organizational perspectives by TCS, Goldman Sachs, and JP Morgan on Hybrid working are discussed.

However, hybrid work arrangements come with certain challenges and complications to be dealt with. The organizations would have to develop norms and policies to maintain optimum productivity and tracking the

Reshaping the Business World Post-COVID-19: Management Strategies for Sustainable Behavior Change. Arvind K. Birdie & Ruchi Joshi, (Eds.)

employees' performances, to attract and retain talent specially at the times such as the Great Resignation 2021, to cater to the changing training needs on leadership and managerial skills to handle hybrid teams, planning for redesigning of work spaces and infrastructure, establishing networking and building effective teams, fostering values and imbibing organizational culture for better employee engagement, providing fair and equitable inclusive work environment, and to tackle issues related to burnout and digital exhaustion and the fear of missing out (FOMO) among employees. Hybrid working can be a game-changer for the future of organizations' functioning and success, in many, if not in all organizations. The organizations must not expect the hybrid work model to evolve by itself. There is no one size fits all model of hybrid work. The time is not to be overly dogmatic and more time is needed to understand the broad contours of flexibility at workplaces. Hybrid work policies need to be developed by gaining from the hybrid work experiences by the employees and the organizations.

2.1 INTRODUCTION

The world of work has drastically changed over time. An ancient era was dominated by manual labor. The invention of electricity led to an industrial revolution which has resulted into use of various machinery and equipment in production, operations, and other types of jobs. The world of work changed further with the invention of the internet resulting into digital revolution. The world economies got connected through globalization. COVID-19 pandemic is also a massive change as the invention of electricity and invention of internet were. This pandemic has immensely affected the world of work and has led to hybrid revolution. The COVID global pandemic is the biggest experiment in the history of work (Gartner Survey, 2021).

The outbreaks of pandemics have impacted human lives. COVID-19 pandemic has also profoundly influenced our orientation, cognitions, and future outlook towards health, work, education, and life in general. Organizations across the globe were able to sustain through remote working during and after the COVID-19 lockdowns. Major economic changes have impacted the world of work and the organizations. Such changes have earlier affected the organizations belonging to specific categories; however, the COVID-19 pandemic has exerted a widespread profound impact on all the organizations. The future of organizations is thought to be heavily influenced by such changes brought by the pandemic.

COVID pandemic has forced the organizations to pass through various phases in which the work patterns and places have changed. During the lockdowns, a majority of the workplaces remain fully closed initially. After that, the organizations shifted to the mandatory work from home (WFH). When the pandemic scenario improved, the organizations were encouraging the employees to WFH due to the restrictions by the governments and the administration. With further improvements in the pandemic scenario, the markets and the organizations opened fully. Such transitions have made the organizations go through a constant change in the working patterns but had made the organizations realize that work can be done and supervised in remote settings as well. Organizations have started envisioning and adopting the hybrid work culture.

One major change in the world of work has been the percentage of people working from homes has substantially increased. The organizations which were earlier resisting the employees working from home had to adopt WFH during lockdowns and thereafter. This sudden unplanned change in the world of work has made the organizations realize that the optimum productivity could be obtained with the WFH model as well. According to International Labor Organization, 7.9% of the global workforce, which is approximately 260 million workers, was engaged in WFH before the COVID-19 pandemic. After the outbreak of the pandemic, WFH employees rose between 20% and 30% of the global workforce in various countries (ILO, 2021).

The World Bank in its report titled 'Who on Earth can work from home' has stated that globally one of every five jobs can be performed from home, although this number is enormously heterogeneous among the countries with different levels of income. In low-income countries, one of every 26 jobs can be performed from home. Poorly educated, temporary workers are more vulnerable to labor-market related changes observed during the pandemic. Educated and better paid workers have more opportunities for work-from-home and can absorb the recent pandemic-related challenges. According to the World Bank's working paper by Sanchez et al. (2020) COVID-19 has created inequalities among countries and between regions within countries.

2.2 CONCEPT OF WORK-FROM-HOME

The concept of work from home (WFH) was introduced by Jack Nilles and his colleagues in the 1970s. They used the term teleworking or telecommuting referring to the use of computer technology and telecommunications to complete the work when away from the routine workplace (Nilles et al., 1976). The features of WFH include the employees work away from their primary or central workplace typically their homes. They use technology, computer, and other electronic devices for interaction and work progress (Allen et al., 2015).

2.3 WORK PATTERNS IN ORGANIZATIONS

Organizations have gone through various changes with time. Organizations have become more oriented toward team-based and collaborative work, are more technologically dependent and are more agile to respond to the changing environment. Responding to the competitive market by preparing appropriate policies, altering the strategies, developing the required skill set among the employees are must for its survival. The organizations, after the COVID-19 pandemic, have undergone a massive unplanned experiment by providing its employees the facility of 'remote working' or 'work from home (WFH).' Across the globe, the organizations continued its functioning, to the extent possible, by its employees working from home.

The use of technology has enabled organizations to provide their employees the facility of flexible work arrangements. The COVID-19 pandemic has further necessitated the flexible work arrangements. Various terms are used synonymously for flexible work arrangements (Joyner, 2021).

Telecommuting refers to the employee working away from the office; one can work from home, a coffee shop, a library, or a co-working space. The work-from-home term is used when the employee is conducting the work from one's home only. Another term, remote working is also used interchangeably for telecommuting; however, the two differ in the notion that a remote worker can operate from anywhere in the world, and not necessarily be in the proximity to the employer's office. Remote working can be a 'working from anywhere' type of work arrangement. Condensed work weeks refer to providing 4-day or 5-day workdays in a week. Flextime is another work arrangement where the employees can choose the time at which they would work and need to complete minimum hours of work or work units for a work-day.

2.4 HYBRID WORKING

Hybrid working refer to a mix of office and home work arrangements where some employees work on the site (office) by being physical present there, some employees work-from-home or remotely. The employees are provided freedom to choose some days for on-site work and some days for work-from-home. A proportion of employees can take turns to be present physically in the office based on their needs, roles, circumstances, and objectives.

The work-from-homes models can be synchronous or asynchronous. In synchronous settings, all the employees work at the same time, although the workplace is different whereas in asynchronous settings, the employees work at different times and at different places too. Asynchronous work pattern does not demand the employees to work at the same time; they can work at their preferred time at their own pace.

2.4.1 OPTIONS FOR ORGANIZATIONS

The 2021 Gartner Hybrid Work HR Leader Survey emphasized breaking the location centric approach to work and redesigning human-centric work model which provide flexibility to employees and empowers and engages them in work. To do so, organizations must dispel certain myths about hybrid work models. Among the myths discussed in the Gartner report are work will get back to normal soon; and employee productivity is less when they work remotely.

This report has suggested three options to the executives in the organizations (Tiffany, 2021):

- **Option 1:** Reject the hybrid environment and force the return of traditional 9–5 schedule. This option risks more employee attrition.
- **Option 2:** Accept the hybrid work environment but continue following the site-centric practices such as meetings as was seen during the COVID pandemic. It would result in more work fatigue among the employees.
- **Option 3:** In the new hybrid work environment, develop a human-centric design which emphasizes learning, innovation, and equity. This option is the future of work, according to the Gartner report.

The past was office-centric (in-person), the present is virtual office-centric (remote work), and the future is a hybrid environment which is to be developed by the organizations.

Overall, various reports discussed so far have indicated that hybrid work is here to stay and will define the post-pandemic workspace.

2.4.2 MODELS OF HYBRID WORKING

Hybrid working is defined as the mixture of both on-site (in-office) working and WFH or remote working in the employees' schedule. Once the organizations decide to adopt hybrid working, there is no one size fits all approach of hybrid working. Various models of hybrid working are available for the organizations to choose from (The advantages of the hybrid work model – UK research data, n.d.):

1. **The At-Will Model:** In this model, the employees are given the option to choose the work arrangement that is best suitable to them on a given day. The employees process a 'work from office' request for day/days to which the organization responds.
2. **The Split-Week Model:** The workweek is divided into two parts WFH and work from office. A week can be split between 2 and 3 days of WFH and 2–3 days of work from the office. The organizations may allocate days of office work looking at the needs of various departments, managers, and projects. As the employees reach to the office at least 2–3 days in a week, it provides opportunities for social interaction and relationship building which are essential for achieving productivity and growth.
3. **Shift Work Model:** The employees work in alternate shifts in this model. The employees can choose to WFH and then in morning or evening shift at the office. The shift duration can be decided by the organization. In comparison to the Split-week model, where WFH and work from office is decided for each week, the shift model can have a fortnight or more or less duration depending upon the organizational functioning and requirements.
4. **Week by Week Model:** In this type of model, the employees work at home and at the office in alternate weeks. For example, if the employees WFH in the first week, the second week would be at the office, this alternation of work-weeks continues.

2.4.3 REMOTE AND HYBRID WORKING STATISTICS

A survey by Gartner, Inc. conducted on 127 company leaders from the field of HR, Legal, Finance, Real Estate revealed that 82% of the leaders were agreeing to allow the employees WFH some of the time, whereas nearly half of the leaders (47%) had intended to allow the employees to WFH for full time (Baker, 2020).

During the transition phase in the world of work during the pandemic, many office-based companies adopted the work-from-home option to sustain its operations and maintain the productivity. The organizations have realized the benefits of the remote working, An HR Consulting Firm, Mercer, conducted a survey in May 2021 on more than 500 employers and found that 70% of the companies were planning to adopt the hybrid work model (Beth Braccio Hering, 2019).

According to Nicholas Bloom, an Economics professor from Stanford University, work-from-home for two days a week optimally would balance collaborative and quiet work and would also benefit from reduced level of stress among the employees as they would need to commute less (Ro, 2020).

Another survey by CIPD conducted in the United Kingdom on 1,046 employers showed that employers have found employees to be productive at WFH settings as well. Around 28% of employers believed the productivity increased, 37% believed there has been no effect on productivity (it means it didn't decline also), whereas 28% of employers reported a decrease in the productivity of employees. Around 70% of the employers planned to include WFH on a regular basis, 45% planned to introduce 'always working from home' models (CIPD, 2020).

A survey by US-based Enterprise Technology Research (ETR) on 1,200 Chief Information Officers (CIOs) has also found that the percentage of remote working employees would double in the year 2021 due to positive productivity trends reported in the year 2020 after the pandemic. The percentage of remote working employees would be double (34.4%) in comparison to the pre-pandemic times (16.4%). Among the respondent information officers, 48.6% reported to find improvements in the levels of productivity after the remote work began whereas 28.7% reported a decline in the level of productivity (Chavez-Dreyfuss, 2020).

The 2021 Work Trends Index, a report based on a study on more than 30,000 people in 31 countries, has also found that the flexible work is here to stay. The report also includes various productivity and labor signals at Microsoft 365 and LinkedIn and the perspectives obtained from experts. Over 70% of the employees want the flexible work option to continue in the future. On the other hand, 65% of the employees are also craving for a more in-person time with other employees/teams. Such contradiction among employees' orientation towards hybrid work is referred to as 'the hybrid paradox' where at one end the employees are willing the flexible work options to continue whereas on the other end, the employees are also missing the in-person time with other employees thereby missing the social interaction, relationships, and workplace chits chats.

The researchers have considered WFH and its facets as important research questions. Brynjolfsson et al. (2020) found that 34.1% of Americans switched to WFH during COVID-19 whereas Dingel & Neiman (2020) concluded that 37% of the jobs in the United States could be performed from home.

Yang et al. (2020) studied the nature of work and WFH possibilities and concluded that WFH is more suitable and beneficial for those occupations that require focused time in large blocks. WFH may not be suitable for those works where people need to collaborate more to accomplish the tasks.

Bloom et al. (2015) conducted a study for a call center company which kept 50% of its workers to WFH arrangement and found that these employees missed fewer days of work and took less break time in between the work which lead to a 13% increase in the number of calls answered.

International Labor Organization (2020) estimated that the share of work-from-home employees will increase in the global workforce after the pandemic. Before COVID-19 pandemic, ILO estimated 7.9% of the global workforce to WFH. ILO says that 18% of the worker population lives in the countries where technology would enable WFH arrangements.

Berg et al. (2020) stated that the scope for WFH differs across the regions. In North America and Western European countries, 30% workforce could WFH, whereas sub-Saharan African and South Asian workers, 6% and 8% of workers can WFH. The Latin American and Eastern European have 23% and 18% workers working from home, respectively.

Estache & Tooth (2020) opined that the COVID-19 virus crisis caused a natural experiment for WFH. Although not all activities and jobs can have possibilities of WFH, however, if the business organizations can look for WFH as a sustainable alternative by developing appropriate strategies and there is a room for WFH in organizations.

2.5 OCCUPATIONS AND WFH

Not all occupations have possibilities of WFH. Occupations differ in various ways, some involve direct dealing with the public, and some occupations cannot be performed without operating machinery and devices on site. Physical presence at the workplace is mandatory in some occupations. The concept of work-from-home, remote working, and hybrid working, is thus, applicable to a limited number of occupations and persons. Even in an organization not all occupations, not all roles and not all levels can WFH.

Baker (2020) also stated that not all occupations can be suitable for WFH arrangements. The National Employment and Wages data by the United States Bureau of Labor Statistics was merged with some other survey data which found that 19.5% of US workforce (total 144.7 million persons employed in US in May 2018) which is approximately 28.2 million people are employed in such occupations where high level of public interaction is required and there is minimal use of computers; hence 19.5% of US workforce cannot work in WFH settings.

Baker (2020) has divided the occupations in four categories based on the level of public interaction and level of computer use:

- **Less Public Interaction, High Computer Use:** This category includes occupations related to finance, technical, architecture, engineering, some administration and some management, etc.
- **More Public Interaction, Less Computer Use:** This category includes occupations such as retail, food service, beauty services (barbers, hairdressers, and manicurists), protective services, etc.
- **Less Public Interaction, Less Computer Use:** This category includes occupations such as production, construction, maintenance, natural resources, transportation, etc.
- **More Public Interaction, High Computer Use:** This category includes occupations such as education, healthcare, sales, some administration and some management, etc.

2.5.1 WORK FLEXIBILITY IN KNOWLEDGE ORGANIZATIONS

Knowledge workers have more opportunities for working from home; however, the employees who are not knowledge workers are also a big chunk of organizations. In order to attract and retain talent, the organizations which develop appropriate strategies for incorporating flexibility of work for the non-knowledge workers would get benefit in the war for talent in the volatile labor market. According to a survey by Gartner, Inc. conducted on 5,000 employees in January 2021, nearly 50% of the employees who are not knowledge workers expressed the need of flexible work arrangements in the work schedules. One-third of non-knowledge workers have flexibility in some area of their work. Another survey by Gartner on 4,264 employees has revealed that 75% of knowledge workers expected more from their organizations, enabling more flexibility in working (Baker & Zuech, 2021).

Although it has been said that certain occupations and jobs are not fitting in with the remote working models; then also Gartner, Inc. suggests certain ways by which the leaders and managers can strategically plan to provide flexibility to the other categories of employees. Rather than classifying jobs that can be performed from home and from the office, the managers should think about the activities in each job that can be performed from home and from the office, this would help in providing flexibility to the other left-out employees as well. It might appear difficult initially, but to attract top talent in the competitive markets, the organizations can exercise this to identify the possibilities. According to Emily Strother, senior principal in the Gartner HR practice, when the flexible work arrangements are provided to the employees, the percentage of high performers increase by 40%.

2.6 RESEARCH ON FUTURE OF WORK

2.6.1 FUTURE IS HYBRID: WORK TREND INDEX 2021

Microsoft's Work Trend Index 2021 report is titled 'The Next Great Disruption is Hybrid Work – Are We Ready?' This report, based on a study of over 30,000 persons in 31 countries in January 2021, has also highlighted that hybrid work is here to stay. As 41% of the global workforce is considering leaving the jobs, 46% of employees are likely to change jobs as the option to work remotely is available. The approach of the organization towards hybrid work and its detailed planning is critical for the organizations to attract and retain talent.

Employees face various challenges while working remotely. Availability of resources is one such issue. Around 42% of the employees have reported that they lack office essentials while working remotely. Every 1 in 10 employees does not have adequate internet connectivity. These problems are there when the employees have worked remotely for more than a year now. Only 46% of the employees, globally, were helped by their employer for their remote work expenses. In China, 78% of employees have been helped by the employer for such needs.

The employees are getting more work fatigue to digital exhaustion. Dr. Mary Donohue from the Digital Wellness Center considered the speed and urgency of virtual mode as the cause of this exhaustion. In person conversations are helpful in understanding the body language, social cues and tone which helps in avoiding digital static that causes work fatigue.

Different categories of employees have different issues while working remotely.

The new employees and Generation Z is facing more challenges as they might have just started their careers, and it is difficult for them to understand and face the workplace challenges. If it would have been a normal work time, not the pandemic, then they would have experienced the joining, training, networking, and building relationships with the fellow co-workers.

According to Riverbed/Aternity Hybrid Work Global Survey 2021 conducted on nearly 1,500 business leaders, almost 85% of the leaders believed that one fourth of the workforce will work in hybrid work settings even after the COVID pandemic is over. In the same survey conducted in the year 2020, the percentage of such business leaders was 53%. However, 94% of the leaders considered hybrid working helpful in retaining top talents and in the competitiveness in the market. In this survey, 88% of the leaders expressed concerns over the digital disparity between in-office and remote workers' home settings that is why 89% of leaders said to invest in technology in the next 12–18 months (McKendrick, 2021).

The survey also identified the barriers that can affect the hybrid working. These barriers include lack of employee motivation and well-being, problems related to technology and its disruption, poor performance due to poor networking among employees/teams, difficulties related to virtual relationship building and issue related to lack of collaboration, increased security risks, etc.

KPMG CEO (2021) outlook survey conducted on 1,325 CEOs across 11 major markets found that hybrid working could be a long term trend for the organizations and it need to revise the policies, strategies, and culture to attract and retain the top talent. The survey also revealed a change in the perception of the CEOs towards the downsizing of the office space. In 2020, 69% of the CEOs were planning to downsize the office space whereas in the survey conducted by KPMG in 2021, only 21% of the CEOs are considering to downsize the office space. However, 51% of CEOs were also looking to invest in shared offices so that they could allow more flexibility to their employees thereby enabling WFH or remote work options. Every 4 out of 10 CEOs are planning to hire remote workers.

The National Association of Software and Services Companies (NASSCOM), India, conducted a survey to understand the future of work outlook for Tech industry in India. Almost 70% of the organizations are looking at hybrid work model even after the pandemic. The survey estimate that 50% of the workforce will return to office at least for up to three days a week by the January 2022. The survey also found that the organizations with employees more than 1,000 employees are more likely to adopt a hybrid working model (NASSCOM Return to Workplace Survey – Evolving Towards Hybrid Operating Model, 2021).

Harris (2021) considers the data on hybrid a mess. The employees don't know what they want. The employees have favored two competing ideas. At one end, they feel isolated and need more social connections but when asked about joining offices back, they would say no. He further emphasizes that two years from now, our definition of hybrid will be much more nuanced and much more evolved.

2.7 BUSINESS LEADERS PERSPECTIVES ON HYBRID WORKING

The best talent available for a specific work might consider the option to remotely work as a deal breaker while choosing an organization for working. Various organizations in some specific sectors have already made provisions for adopting hybrid and remote working a part of their organizational policies.

Microsoft's Chief People Officer Kathleen Hogan says:

> *"I definitely would be thoughtful about if you can't enable hybrid, having a really good reason why it's really important for the people to be in the office 9–5" Hogan has further added "we will offer as possible to support individual work styles, while balancing business needs and ensuring we live our culture."*

Microsoft CEO Satya Nadella says:

> *"Employees expectations are changing and we will need to define productivity much more broadly inclusive of collaboration, learning, and well-being... All this needs to be done with flexibility in when, where, and how people work."*

Google's Sunder Pichai has said:

> *"The future of work at Google is flexibility" "We are moving to hybrid work week with most Googlers in the office approximately 3-days a week."*

Google expects that 20% of its employees will permanently WFH whereas 60% will work on a hybrid mode. Google has also provided its employees to work from anywhere for 'at least 4 weeks in an year' after getting their manager's approval.

Mark Zuckerberg from Facebook has predicted that 50% of its employees will be working remotely in the next 5–10 years. Facebook Vice President of People Growth Barry Harrington has, however, expressed the possibility of variability in terms of pay based on the local cost of labor in the market from where employees work. Twitter has also shown its commitment for a full remote optional future. They have declared a policy of 'Get to Yes' on all relocations requests received from the employees (What Top CEOs are Saying about the Future of Hybrid Work, 2021).

Although organizations like Goldman Sachs and JP Morgan have expressed varying opinions on the possibility of hybrid work. Goldman Sachs CEO David Solomon called remote work not a new normal and an aberration. Solomon says, "I do think for a business like ours which is an innovative, collaborative apprenticeship culture, this is not ideal."

These organizations have expressed concerns over the notion that they might lose business deals due to lack of interpersonal touch. Lack of such one-to-one interactions while the organizations are vouching for and finalizing business deals is strong argument given against the hybrid work culture. An important though to consider in such situations is that are such deals only dependent on one to one interactions? Do certain other factors such as quality of product and/or services, timely delivery, brand reputation, organizational branding and trustworthiness matter? If the organizations which are trying to get business deals are at par on the above parameters, one to one interaction might have its impact.

TCS' 25-25 Model: A new work mode adopted by Tata Consultancy Services (TCS) has rendered the age-old concepts once considered necessary for optimal productivity. TCS has changed its work model by reducing location dependency, fixed working hours, the necessity of employees to be present in the office (presenteeism), and the need to high-touch governance. TCS has planned to develop location independent work models which will allow the organization to take the maximum advantage of the talent ecosystem. Almost 95% of the TCS employees have worked remotely and have managed 23,000 projects during the pandemic. Although TCS had initially asked its employees to return back to their base locations after the COVID restrictions were lifted, but eventually one fourth of its workforce will work from the office by the year 2025. TCS has announced its 25 by 25 vision by 2025. It means that only 75% of TCS workforce will WFH by the year 2025. And the 25% employees working from office will also work only 25% of their time at the office (TCS SBWS Secure Borderless Workspace™ paves the way to make virtual workplaces the new normal, n.d.).

2.7.1 FLEXIBILITY VS. RADICAL FLEXIBILITY

Different people have different hours of the day when they are most productive. The concept of 'radical flexibility' has been explained by the Gartner Inc., which allows the employees to choose when and where they would work. Radical flexibility offers more flexible work arrangements to the employees in comparison to the 'flexibility.' In radical flexibility, the employees can work on any days of the week and can customize their work hours themselves. The employees get the opportunity to work at their best hours and are able to manage their personal and family responsibilities which promote work-life balance for them. Once the employees have met with the client and teams and have understood their responsibilities, they need to communicate their work schedule to their managers and colleagues. In one organization, Saam Architecture, radical flexibility is provided to the employees. They can commute at off-hours also, if needed. The employees cover a 32-hour/week schedule through a shared office calendar in such a way that it support both the clients and the employee oneself (Nicklaus, 2020).

2.7.2 CHALLENGES AND RECOMMENDATIONS FOR HYBRID WORKING

Various organizations and business leaders have advocated the possibilities of hybrid working in the future; however, the hybrid culture also poses certain challenges for which the organizations need to develop appropriate strategic interventions. Adopting hybrid work patterns brings its own challenges and complications to deal with. Some of these have been discussed along with the recommendations to deal with these issues.

2.7.2.1 TALENT ATTRACTION AND RETENTION – THE GREAT RESIGNATION 2021

According to the US Bureau of Labor Statistics, the trend of employees leaving their jobs is consistent in the USA. Approximately 4 million Americans quit their jobs in April 2021 which result into 19.9 opens job at the end of this month. The phenomenon of a rapidly growing number of Americans leaving their jobs is referred to as 'the great resignation' or 'the big quit.'

Cook (2021) conducted an extensive study in which the data of more than 9 million employees' records from more than 4,000 companies was explored. It was found that the rate of resignation was high among the mid-career employees (between 30- and 45 years of age). It might be due to the demand of more experienced people in comparison to new or untrained workers. The employees could also leave their jobs after the pandemic times high workload, and the pent-up resignations might have found the right opportunity. The study also found that resignations were highest in tech and healthcare industries. Although some business leaders and researchers have considered 'the great resignation' a temporary state; however the organizations must consider adopting various measures to attract and retain talent. When a trained employee leaves the job, the remaining team or employees feel the gap of resources and skill-set. This gap impacts on the quality of work and the timely delivery of work done. The organizations were already suffering from high rates of attrition even before the pandemic.

Cook (2021) recommends the organizations to quantify the impact of resignations. The data of resignation need to be categorized into voluntary resignations and layoffs/firings. Then the root causes of resignations need to be understood, whether it is due to the issues related to pay, compensation, promotion-related issues, or it is due to the interpersonal conflicts among teams/boss, policies, and appraisal-related issues. The organizations must develop a data-driven employee retention strategy which addresses the root causes identified.

Mark Dixon from International Workplace Group has, referring to the Gartner Inc. survey, said that 85% of the employees have said that if their company forces them to come into the office five days a week, they would start looking for another job. If the organizations want to keep their talent, the talent is telling to provide an office space near their homes or allow them to WFH. However, Dixon also emphasizes the importance of employees being interacting in person as it fosters a sense of belonging and a feeling of purpose. If the organizations do not bring people together, they would become digital nomads and would lose the organization culture. The organizations need to develop a number of office spaces that are close to the employees. People did not work hard, got educated to spend 3–4 hours each day on commuting (Dixon, 2021).

Premkumar (2021) talked about a direct link between work models and resignations. She emphasizes that COVID has un-tethered us (employees) from routine, it has given us a chance to think about what we really want in our careers. Such shifts in perceptions and cognitions have caused the great resignation and high attrition across the organizations globally.

Balasubramanian (2021) considers multiple factors responsible for the great resignation. Many employees have made their life decisions while working from home during the pandemic. They have moved permanently move to their native places to stay with the families and they want more of it now. The resignations can also be due to the pent up career choices which have piled up during the COVID phases. The organizations have also realized that work can be done through offshore delivery; hence the new jobs have been designed keeping the new trends in mind which has attracted the talent pool.

On the other hand, some viewpoints considering the great resignation a temporary state have look at the rising wages year over year, which is a sign of an empowered workforce trying to fetch a higher pay. The high quit rate is also looked at as a measure of economic confidence due to which employees voluntarily leave their current jobs and accept newer roles. The great resignation data need to be understood in the context of the hiring data by the same Labor department. According to this data, the hiring is progressing more than the resignations. Around 6.5 million people have been hired in September 2021, which is 2 million more than those leaving their jobs (Cox, 2021).

2.7.2.2 ACCESS TO TALENT: GIG ECONOMY

The hybrid work policy or a fully remote working policy can enable organizations to hire talent available in distant locations. Another advantage, from the organizational perspective, is that the talent can be hired for a short term or on a freelance basis which also reduces organizational costs. The type of labor market which is characterized by the prevalence of contractual work that is for short term or on freelance basis is known as Gig economy. The independent workers are hired with short term commitments. The availability of high demand and specialized knowledge and skills are accessible in Gig economy. It not only provides flexibility to the recruiters but is also a cost-effective option. From the employees' point of view, it gives them the freedom to set their own working agenda. A fully remote or hybrid working models are more suitable to such employees preferring to work on a freelance or short term basis.

2.7.2.3 EMPLOYEE PRODUCTIVITY AND MONITORING

The evaluation of productivity is a greater challenge when shifting to the hybrid work pattern. To evaluate the productivity of employees in the new set-up, organizations need to devise ways. Organizations need to address the challenges arising due to the lack of results and the issues related to determining accountability in poor performing teams.

In a hybrid work environment, the organizations have evolved strategies based on data driven processes. The software can be helpful for finding the relevant information which the organization can analyze to monitor the work progress. The use of employee information and analytics is thus highly useful in the hybrid work environment. Tracking of work computers of an employee while working from home, the monitoring of work communications, emails, meetings, and text messages can be the sources for obtaining the requisite information about employees work involvement.

2.7.2.4 LEADERSHIP AND MANAGERIAL SKILLS

Maintaining a productive hybrid work environment would be a challenge for the leaders. The traditional ways of leading teams in the in-office setup might not fulfill the changing requirements in the hybrid work setup. The leaders could look for setting up rules for engagement with their team members. The expectations of the performance, deadlines, and targets should be clearly discussed in an open environment of trust. The team can remain connected with each other through informal ways also to build strong virtual relationships. A leader willing to walk a distance with the team members during tough times could foster work performance (Guest, 2021).

Work from anywhere is not a simple addition to the organization structure. The employees, the managers and leaders are affected in various aspects. The employees might not feel that sense of 'psychological safety' to trust a team's decision in the absence of one's physical presence in the office. The employees can also get 'fear of missing out' (FOMO) in the virtual meetings if left unheard. The managers would not automatically know how to handle such hybrid teams. Skill training is required to effectively lead such teams for both achieving optimum productivity and for equitable and fair treatment to all (Premkumar, 2021).

Satya Nadella from Microsoft favors an organic process approach to develop norms for hybrid work related structural changes. According to him, the two megatrends that are changing the organizations are the hybrid work arrangement and the great reshuffle of employees. The managers know their teams well, what are their needs, preferences, performance, life situations such as commuting time, child care or other responsibilities. The managers need to be trained on developing the norms that work best for their teams. Such process will enable organizations to develop a broader framework for incorporating appropriate changes in both organizational structure and policies that foster optimum performance and employees' well-being in the hybrid work setup (Nadella, 2021).

2.7.2.5 NETWORKING AND TEAM BUILDING

Remote working has led people to lose human connections that were formed through informal impromptu chit chats and in-person coffee-breaks conversations. The lack of such interactions leads to lack of formation of interpersonal relationships among team members and leaders. Strong interpersonal relationships lead to better functioning of teams and thus lead to optimum productivity. Thus hybrid work, thus, poses challenges to the organizational structure, communication, and interpersonal relationships.

The team leaders can be trained to provide the leadership that is suitable for managing teams in a hybrid work setting. The skills needed to manage an in-office team and a hybrid team differs. A strategic plan for the requisite skills development by the Human Resource Developments is the key here in understanding the changing organizational structures and its impact, the changing training needs and skill gaps for the employees and leaders. Such analysis would lead to designing an appropriate training intervention which addresses the challenges posed by the hybrid work environment.

Business leaders can plan different ways through which the emotional connections among the hybrid teams could be fostered. The employees could be discussing, other than the defined work deadlines and scheduled tasks, about the shared beliefs and values of the organization, what the organizational culture, how the organization has scaled such heights, how various brands have grown, what were the obstacles, how it overcome these obstacles over the years, etc. When employees discuss such issues, it creates a sense of belongingness to the organization for which they are serving. When employees know what they are doing, why they were doing so, it forms a sense of purpose which is connecting element for the employees. Belonging through the sense of purpose is a factor that can promote the growth of the organization.

2.7.2.6 SWITCHING COST

To develop the requisite infrastructure and digitization, the organizations would bear the additional cost. The expenditure on employee wellness programs would also increase. The organizations might need to organize special training programs to accommodate employees with the changing needs of the hybrid world. However, the employees can also save financial resources as they might need to pay less for office space. Some surveys have also shown that the employees are ready for a pay-cut to trade for a relocation request. The Gartner Inc. survey 2020 identified that 48% of the employees were ready to relocate to other locations. Among these employees, 62% were ready to accept a pay-cut also (Wiles, 2021).

Vaughan Klein from Cisco, a tech company, states that their company has saved $500 m while working on hybrid mode over the last five years. Cisco was reducing its real estate footprints before the COVID pandemic also. The savings on the real estate were invested on other heads (Calculating the P&L of hybrid work, n.d.).

2.7.2.7 ORGANIZATIONAL CULTURE AND VALUES

A young employee who has just joined in does not get to thoroughly the organizational culture and values due to lack of in-person interactions with the people and the organizational environment. That lack of personal touch could be a deciding factor in remote settings.

To inculcate the organizational values among the employees and to lighten the work environment, some activities and get-togethers can be planned even in the hybrid working environment. At this time, the employees would not talk about the work but would be informally interacting through various activities, parties, award functions, welcoming the new joiners, to name a few. Such virtual informal interactions would not only create better understanding employees but would also improve communication and networking among them. By indulging in such activities, the lack of those 'buffer moments' of interaction the employees miss in the hybrid work environment can also be overcome.

A hybrid work model is to be based on the policies which provide equity and fairness to all employees. The policies for a hybrid work culture must be framed considering the sense of equity to all employees in allocation of resources, work schedules, appraisals, increments, promotions, etc. The faces which are more visible at the workplace have more chances for better performance appraisal; hence a policy which transparently assesses the work performance and productivity of an employee irrespective of the place where an employee works is recommended for hybrid organizations (How Leaders Should Approach Today's New Hybrid Workforce, n.d.) When employees perceive equity and fairness at the workplace policies, their engagement with the organizations and work performance is improved.

2.7.2.8 FAIRNESS AND EQUITY

Both the employees working from home and the employees working from offices have their own sets of advantages and disadvantages. WFH employees have the advantage of saving time, money, and hassles on commuting to the workplace. In-office employees might have the advantage of visibility in the eyes of the boss which could lead to faster career advancement. A performance appraisal based on the actual deliverables would thus be helpful. The decisions related to providing flexibility in work should be based on a set of principles designed in consultation with the employees or with their representatives (Brower, 2021).

The organizations which plan to adopt the hybrid work model for a long-term must invest sufficient time and resource to research the employees' perspective not only for framing the hybrid work policies but also for the periodic evaluation of the policies through timely employees' feedback on various facets of the policies and its effectiveness, acceptance, and fairness in implementation. A policy well planned has to be well implemented as such otherwise it is only on papers. The practice of assessment and re-assessment of the hybrid work policies will result in developing a long-term acceptable hybrid work culture through the inculcation of fair principles and norms. Although some categories of jobs are more pro to the hybrid work culture; however the organizations should explore the option of providing flexibility in work in an equitable and transparent manner to all possible categories of jobs.

According to Rosie Campbell from King's College, London, who reviewed past research trends on WFH, the remote-workers have more chances of not getting promoted. This is mainly due to the flexible stigma which is explained as the negative perception of those who are working from home. These employees have more chances of facing discrimination and negative outcomes at the workplace. Thus, the organizations which are planning to adopt a hybrid work model must consider treating all employees with fairness and equity (Jacobs, 2021).

2.7.2.9 INCLUSIVE WORK ENVIRONMENT

The employees, who were working remotely, before the pandemic, have felt left out as an outlier in the meetings conducted in the office. The focus of the offline attendees would be more on those attending offline.

As the pandemic has forced a majority to attend the meetings online, the employees' earlier feeling left out felt inclusive now where the online attendees have increased. Hybrid work pattern could prove to be a risk for running two parallel organizational cultures. One cohort of employees might have more in-person interaction and collaboration, whereas the other pool of employees might feel that their work is not properly recognized by their leaders and/or teams. Such employees might not engage fully with the organizational objectives due to feeling isolated and lesser sense of belongingness with the organization.

2.7.2.10 EMPLOYEE BURNOUT

The employees working remotely during the pandemic have provided their feedback in various survey and have said that they used to feel as if the workday does not end. Their work schedule was challenged by the blurred boundaries between work and personal life in remote working. The companies like Airtable have adopted measures such as 'Recharge days' which gave an additional day-off every quarter to their employees. Everyone did this unplugging of the work at the same time. Airtable also implemented 'Focus Weeks' in which everyone was encouraged to cancel recurring meetings and focus on prioritizing more critical tasks. In the survey, Airtable found that 97% of the employees were satisfied with the 'Recharge Days,' 90% were satisfied with 'Focus Weeks.' These initiatives have led to the reduction in digital fatigue caused due to long-meetings and video calls (Jackman, 2021).

The leaders also need to consider appropriate communication networks among the team members and develop appropriate schedules so that unnecessary online meetings can be avoided. Sometimes a three to five minute conversation at the in-office environment can result into a 30-minute video meeting in the remote working which causes digital fatigue and burnout.

Humans like routines whereas in hybrid working, the employees need to work 2–3 days from the office and 2–3 days from home. The employees need to switch their roles and routines as per the changing work patterns. Not only the employers have to bear the switching cost of the hybrid work by changing the work infrastructure, but the employees also need to bear this switching cost for their mental make-up. According to Premkumar (2021), the hybrid working can cause two-three Monday Morning Blues to the employees. Hybrid working is like asking somebody to play a test match in the first inning, a T-20 in the second inning and then a test match again in the third inning. This switching cost can affect both the mental health and productivity of the employees.

2.7.2.11 DIGITAL EXHAUSTION

As the time and intensity of digital interaction has increased for the employees while working from home, the meetings, communications, submissions are happening through digital technology. The rise in human-machine interaction has led to digital exhaustion among the employees. Microsoft's Work Trend Index 2021 has also identified digital exhaustion as a major challenge in the hybrid work world. The digital fatigue and burnout causes employees at their health, emotions, work performance, dynamism, creativity, and innovation. Microsoft has partnered with some organizations that would help its employees deal with digital exhaustion through mindfulness and meditation. As the organizations are investing in technology, infrastructure, and creating WFH setups, investment on dealing with the mental health related issues among the employees also needs attention. Both physical and mental health are strongly associated with the employee well-being and performance.

2.7.2.12 ROLE CONFUSION

An employee needs to attend to various household needs simultaneously while working at home. Attending to visitors, managing household chores, kids, their homework can create role confusion. As there is no strict time gap or place constraint, it becomes difficult to switch to different roles of worker, parent, partner, and housekeeper (Aarons-Mele, 2021).

2.8 CONCLUSION

The hybrid working is here to stay. The outlook of business organizations and leaders towards the remote work has changed. The advantages and disadvantages of remote working have also been thoroughly understood in this massive unplanned experiment in the form of COVID pandemic. Some categories of organizations, if not all, will need to adapt hybrid working in its policy to sustain and improve productivity. All organizations are not suitable for hybrid work patterns.

The viewpoints favoring both the hybrid work and office work have been expressed by the organizations; however, the success of an organization depends on how it orients itself to the changing work environment and is able to capture the best talent available in the world for its optimum advantage. Hybrid work model offers not only offers flexibility but also empowers the employees so that they can work full to their strengths. A single and similar model of hybrid work might not work for all organizations. The organizations would need to develop a working model based on its own analysis. As Satya Nadella from Microsoft says that this is not the time to be overly dogmatic, let the new work norms settle and then only we would be able to understand the broad contours of flexibility. Time will tell us that how in the long run; the new work arrangements would emerge. One thing is for sure is that the people looking for work-life balance would be attracted towards the organizations that provide hybrid work options to its employees.

ACKNOWLEDGMENTS

The author acknowledges the contributions of the editors of this book Dr. Arvind Gagan and Dr. Ruchi Joshi for their review inputs and suggestions. The author also acknowledges the contributions of the researchers and business leaders cited in this chapter for their perspectives on hybrid work which have been immensely useful.

KEYWORDS

- **chief information officers**
- **enterprise technology research**
- **fear of missing out**
- **flexibility**
- **future of work**
- **hybrid working**
- **organizational policy**
- **work from home**

REFERENCES

(2021). *KPMG 2021 CEO Outlook* – KPMG Global. Retrieved from: KPMG website: https://home.kpmg/xx/en/home/insights/2021/08/kpmg-2021-ceo-outlook.html (accessed on 7 February 2023).

(2021). *NASSCOM Return to Workplace Survey – Evolving Towards Hybrid Operating Model.* Retrieved from: NASSCOM website: https://nasscom.in/knowledge-center/publications/nasscom-return-workplace-survey-evolving-towards-hybrid-operating (accessed on 7 February 2023).

(2021). *What Top CEOs are Saying About the Future of Hybrid Work.* Retrieved from: Align website: https://aligntoday.com/blog/ceos-on-hybrid-work/ (accessed on 7 February 2023).

(n.d.). *The Advantages of the Hybrid Work Model – UK Research Data.* Retrieved from: Hibob website: https://www.hibob.com/guides/hybrid-working-model-advantages/ (accessed on 7 February 2023).

(n.d.). *Work Trend Index: Microsoft's Latest Research on the Ways We Work.* Retrieved from: www.microsoft.com Website: https://www.microsoft.com/en-us/worklab/work-trend-index (accessed on 7 February 2023).

Aarons-Mele, M., (2021). *Anxious ABOUT Hybrid Work? Ask Yourself These 4 Questions.* Retrieved from: Harvard Business Review website: https://hbr.org/2021/06/anxious-about-hybrid-work-ask-yourself-these-4-questions (accessed on 7 February 2023).

Allen, T. D., Golden, T. D., & Shockley, K. M., (2015). How effective is telecommuting? Assessing the status of our scientific findings. *Psychol. Sci. Public Interest, 16*(2), 40–68.

Baker, M. G., (2020). Who cannot work from home? Characterizing occupations facing increased risk during the COVID-19 pandemic using 2018 BLS data. *American Journal of Public Health.* doi: 10.2105/AJPH.2020.305738.

Baker, M., & Zuech, T., (2021). *Gartner HR Survey Finds 60% of Non-Knowledge Workers Want Their Organization to Provide More Flexibility.* Retrieved from: Gartner website: https://www.gartner.com/en/newsroom/press-releases/08-26-21-gartner-hr-survey-finds-sixty-percent-of-non-knowledge-workers-want-their-organization-to-provide-more-flexibility (accessed on 7 February 2023).

Baker, M., (2020). *Gartner Survey Reveals 82% of Company Leaders Plan to Allow Employees to Work Remotely Some of the Time.* Retrieved from: Gartner website: https://www.gartner.com/en/newsroom/press-releases/2020-07-14-gartner-survey-reveals-82-percent-of-company-leaders-plan-to-allow-employees-to-work-remotely-some-of-the-time (accessed on 7 February 2023).

Balasubramanian, A., (2021). *Are we Evolving Towards Hybrid Work Model?* FYI (S. Mehrotra Kapoor, Interviewer). Retrieved from: https://www.youtube.com/watch?v=Aq0c_kiRmgw (accessed on 7 February 2023).

Berg, J., Bonnet, F., & Soares, S., (2020). *Working from Home: Estimating the Worldwide Potential.* CEPR-Vox, https://voxeu.org/article/working-home-estimating-worldwide-potential (accessed on 7 February 2023).

Beth, B. H., (2019). *Remote Work Statistics for 2019: Shifting Norms and Expectations.* Retrieved from: FlexJobs Job Search Tips and Blog website: https://www.flexjobs.com/blog/post/remote-work-statistics/ (accessed on 7 February 2023).

Bloom, N., (2021). *Don't Let Employees Pick Their WFH Days*. Retrieved from: Harvard Business Review website: https://hbr.org/2021/05/dont-let-employees-pick-their-wfh-days (accessed on 7 February 2023).

Bloom, N., Liang, J., Roberts, J., & Ying, Z. J., (2015). Does working from home work? Evidence from a Chinese experiment. *The Quarterly Journal of Economics, 130*(1), 165–218. doi: https://doi.org/10.1093/qje/qju032.

Brower, T., (2021). *Hybrid Work Must Be Fair: 8 Ways to Address the Equity Challenge*. Retrieved from: Forbes website: https://www.forbes.com/sites/tracybrower/2021/06/13/hybrid-work-must-be-fair-8-ways-to-address-the-equity-challenge/?sh=454b18577b39 (accessed on 7 February 2023).

Brynjolfsson, E., Horton, J., Ozimek, A., Rock, D., Sharma, G., & Hong-Yi, T. Y., (2020). *COVID-19 and Remote Work: An Early Look at US Data*. National Bureau of Economic Research, Cambridge, MA. doi: https://doi.org/10.3386/w27344.

Calculating the P&L of Hybrid Work, (n.d.). BBC Future. Retrieved from: https://www.bbc.com/storyworks/future/the-hybrid-economy/calculating-the-pl-of-hybrid-work (accessed on 7 February 2023).

Chavez-Dreyfuss, G., (2020). *Permanently Remote Workers Seen Doubling in 2021 Due to Pandemic Productivity: Survey*. Reuters. Retrieved from: https://www.reuters.com/article/us-health-coronavirus-technology/permanently-remote-workers-seen-doubling-in-2021-due-to-pandemic-productivity-survey-idUSKBN2772P0 (accessed on 7 February 2023).

CIPD, (2020). *Home Working Set to More Than Double Compared to Pre-Pandemic Levels Once Crisis is Over*. Retrieved from: CIPD website: https://www.cipd.co.uk/about/media/press/home-working-increases (accessed on 7 February 2023).

Cook, I., (2021). *Who Is Driving the Great Resignation?* Retrieved from: Harvard Business Review website: https://hbr.org/2021/09/who-is-driving-the-great-resignation (accessed on 7 February 2023).

Cox, J., (2021). *Millions of People Quit Their Jobs in the "Great Resignation." Here is Why it May Not Last Long*. Retrieved from: CNBC website: https://www.cnbc.com/2021/11/18/why-the-great-resignation-may-not-last-very-long.html (accessed on 7 February 2023).

Dingel, J., & Neiman, B., (2020). *How Many Jobs Can be Done at Home?* National Bureau of Economic Research, Cambridge, MA. doi: https://doi.org/10.3386/w26948.

Dixon, M., (2021). *Why The Future of Work is Hybrid Work: Insights from the CEO of IWG*, (J. Morgan, Interviewer). Retrieved from: https://www.youtube.com/watch?v=lm3VMw4HPLE (accessed on 7 February 2023).

Estache, A., & Tooth, S., (2020). *On the Scope for Work-from-home in High and Upper Middle-income Countries*. ECARES working paper, 2020-46.

Guest, S. A. P., (2021). *SAP BrandVoice: From Remote to Hybrid Working: The Role of Leadership*. Retrieved from: Forbes website: https://www.forbes.com/sites/sap/2021/09/28/from-remote-to-hybrid-working-the-role-of-leadership/?sh=6c62646542d2 (accessed on 7 February 2023).

Harris, G., (2021). *What Does the Future of Hybrid Working Look Like?* (D. Green, Interviewer). Retrieved from: https://www.youtube.com/watch?v=ArAwV43k4_8 (accessed on 7 February 2023).

How Leaders Should Approach Today's New Hybrid Workforce, (n.d.). Retrieved from: CCL website: https://www.ccl.org/articles/leading-effectively-articles/how-leaders-should-approach-todays-new-hybrid-workforce/ (accessed on 7 February 2023).

ILO, (2020). *Working from Home: Estimating the Worldwide Potential*. ILO policy brief.

ILO, (2021). *Working From Home: From Invisibility to Decent Work.* [online] Available at: https://www.ilo.org/wcmsp5/groups/public/---ed_protect/---protrav/---travail/documents/publication/wcms_765806.pdf (accessed on 7 February 2023).

Jackman, J., (2021). *What Chief People Officers Need to Rethink for 2022*. Retrieved from: Protocol — The people, power and politics of tech website: https://www.protocol.com/workplace/airtable-radical-flexibility (accessed on 7 February 2023).

Jacobs, E., (2021). *Hybrid Working: Fomo and the Danger of Office Cliques*. Retrieved from: The Irish Times website: https://www.irishtimes.com/business/work/hybrid-working-fomo-and-the-danger-of-office-cliques-1.4568545 (accessed on 7 February 2023).

Joyner, S., (2021). *8 Types of Flexible Work Arrangements Explained*. Retrieved from: ViewSonic Library website: https://www.viewsonic.com/library/business/8-types-of-flexible-work-arrangements-explained/ (accessed on 7 February 2023).

McKendrick, J., (2021). *Business Leaders are Changing Their Minds About Hybrid and Remote Work*. Retrieved from: Forbes website: https://www.forbes.com/sites/joemckendrick/2021/11/07/business-leaders-are-changing-their-minds-about-hybrid-and-remote-work/?sh=6c469add4c86 (accessed on 7 February 2023).

Nadella, S., (2021). *Microsoft's CEO on the Metaverse and Flexible Work*. A. Ignatius, Interviewer. Retrieved from: https://www.youtube.com/watch?v=FRztx0wVuPA (accessed on 7 February 2023).

Nicklaus, D., (2020). *Exploring Remote Work and "Radical Flexibility" in Architecture with Diana Nicklaus*. Archinect, Interviewer. Retrieved from: https://saam-arch.com/culture/exploring-remote-work-and-radical-flexibility/ (accessed on 7 February 2023).

Nilles, J. M., Carlson, F. R., Gray, P., & Hanneman, G., (1976). Telecommuting-an alternative to urban transportation congestion. *IEEE Trans. Syst. Man. Cybern., 2*, 77–84.

Premkumar, D., (2021). *Are We Evolving Towards Hybrid Work Model?* FYI. S. Mehrotra Kapoor, Interviewer. Retrieved from: https://www.youtube.com/watch?v=Aq0c_kiRmgw (accessed on 7 February 2023).

Ro, C., (2020). *Why the Future of Work Might be "Hybrid."* Retrieved from: www.bbc.com website: https://www.bbc.com/worklife/article/20200824-why-the-future-of-work-might-be-hybrid (accessed on 7 February 2023).

Sanchez, D. G., et al., (2020). *Who on Earth Can Work From Home?* Policy Research Working Paper 9347 World Bank.

TCS SBWS Secure Borderless Workspace™ Paves the Way to Make Virtual Workplaces the New Normal, (n.d.). Retrieved from: www.tcs.com website: https://www.tcs.com/tcs-secure-borderless-workspaces-shows-the-way (accessed on 7 February 2023).

Tiffany, F., (2021). *Three Types of Hybrid Work Models (and the Pros and Cons of Each)*. Retrieved from: Envoy website: https://envoy.com/blog/3-types-of-hybrid-work-models/ (accessed on 7 February 2023).

Wiles, J., (2021). *5 Data Insights for CFOs on Remote Work Costs*. Retrieved from: Gartner website: https://www.gartner.com/smarterwithgartner/how-remote-and-hybrid-work-models-affect-costs-5-data-insights-for-cfos (accessed on 7 February 2023).

Yang, L., Jaffe, S., Holtz, D., Suri, S., Sinha, S., Weston, J., ... & Teevan, J. (2020). How work from home affects collaboration: A large-scale study of information workers in a natural experiment during COVID-19. *arXiv preprint arXiv:2007.15584*.

CHAPTER 3

Mapping the Nature of Workplace Incivility in Post-COVID Work Culture

SUYESHA SINGH[1] and RIMJHIM[2]

[1]*Assistant Professor, Department of Psychology, Manipal University Jaipur, Rajasthan, India*

[2]*Mphil Clinical Psychology, Department of Psychiatry, Government Medical College and Hospital, Chandigarh*

ABSTRACT

Workplace incivility has been treated as a lenient and subjugated form of workplace bullying or aggression and has lacked potent action required to curb it in the organizational field. However, the research backing it has systematically and consistently emphasized on the significance of the same on both the productivity and the health of the employees. Workplace incivility not only impacts the employees, but the organization as a whole, leading to gross loss and corporate lawsuits. The psychological health of the workers suffers leading to poor job satisfaction and low work commitment. Post-COVID literature has brought to light an increase in incidents of incivility in workplaces, fueled by the uncertain nature of the situation. The technological advancements do not help either by providing instant communication and open channels, keeping the employees open to workplace environment at all times. However, the studies in the same domain remain scattered, and lack a proper model for the circumstances. This chapter is an attempt at formulating the same by preparing a systematic review of significant and recent studies and figuring out the antecedents and relational outcomes of Workplace incivility.

Reshaping the Business World Post-COVID-19: Management Strategies for Sustainable Behavior Change. Arvind K. Birdie & Ruchi Joshi, (Eds.)

This chapter aims at mapping the research done on Workplace incivility in recent post-COVID times and highlighting the spiraling nature of the same. The impact of singular incidents of workplace incivility ends up developing a whole culture of aggressive environment. It could have short-term and long-term effects. The chapter included a meta-analysis which involved a systematic review of the relevant literature, followed by the careful plotting of the results into a summarized estimate.

- **Implications:** The chapter establishes a spiraling and networking nature of workplace incivility and other acts of micro-aggression, which might make a workspace uninhabitable, and hence bring down the entire organization. The financial and psychological outcomes of the same have been mapped using the previous research, likewise the antecedents have been pondered upon as well. The environmental and organizational antecedents both impact the work culture, and an action plan to curb the same has been suggested. The research also presents that the loss faced due to outcomes of workplace incivility far proceeds the loss which the organization might have to face by dealing with these antecedents. Instead, such actions on the part of the organizations and leaders have been showcased to promote positive workspace culture leading to financial gains.

3.1 INTRODUCTION

3.1.1 WORKPLACE INCIVILITY

Workplace Incivility has, over time, become a significant study topic over the past years, especially since the corporate work sector has increased manifold due to globalization and capitalization. The concept of Workplace Incivility refers to the workplace behavior which has no regard for another person or does not follow the civil norms of the organization. It has been defined consistently (Pearson et al., 2001; Cortina et al., 2001; Blau & Anderson, 2005; Lewis & Malecha, 2011; Reich & Hershcovis, 2014; Schilpzand et al., 2016; Sharma & Singh, 2016; Anju & Raaj, 2019; Kanitha & Naik, 2020; Hulsheger et al., 2020; Greenslade, 2021) over the decades as low intensity deviant behavior with ambiguous intent to harm the target, in violation of workplace norms for mutual respect. This basic definition has ruled many research ideas and defined theoretical framework for experimental studies.

This definition has been used as the foundation stone for many studies and methods of study when it comes to researching workplace incivility. Another definition employed often (Megeirhi et al., 2020) describes Workplace Incivility as employees' lack of regard for one another (Pearson & Porath, 2005). Other researchers have often adapted these definitions to suit their circumstantial and populace needs.

Other ways to define workplace incivility has been through examples of uncivil behavior in workplace, which mostly include angry outbursts, sending rude e-mails, neglecting co-workers, excluding co-workers, disrupting others, name-calling, inability to empathize, damaging co-workers' reputation, shouting, inappropriate jokes, insult or destructive criticism, invasion of privacy, threatening, intimidation, withholding information, unfair treatment and inappropriate language (Pearson et al., 2005; Felbinger, 2008; Samosh, 2019). Post-COVID-19 research (Brass, 2021) has included the decorum for virtual contact, especially E-mail communication manners. To understand the workplace incivility in action, one needs to identify these behaviors and be able to mark them when noticed.

This research chapter focuses on combining the workplace incivility researches to figure out the spiraling impact of its antecedents and outcomes. This spiral is showcased in reference to the present COVID crisis. Now, as the workplaces are opening again, there is an atmosphere of unsurety regarding the work culture and health safety. After being in a work-from-home mode, it will be interesting to observe how the workplace bounces back to its previous momentum. Also, this new beginning provides a great window of opportunity for setting up new workplace precedents and snip harmful workplace behaviors before they fester and the impact the harmony of the organizations.

3.2 METHODOLOGY

The research chapter was written utilizing the vast literature present in the various databases, pertaining to workplace incivility, in both the pre-COVID-19 times and the post-COVID-19 times. The data collection included finding, filtering, and analyzing the research papers and studies found on the internet through google scholar. The research were found using some efficient keywords such as "Workplace Incivility," "Workplace Incivility and COVID," "Workplace Incivility Spiral," "Workplace Culture pre and post-COVID," 'Cyber workplace incivility,' "Virtual work culture," "Workplace Incivility AND Antecedents," and "Workplace Incivility AND Outcomes." It was ensured that major studies about different aspects of workplace incivility be included, and the research which replicated other studies and their results were filtered. The studies which didn't include the variable workplace incivility, and instead had some variant version of the same were discarded. Only that research which was published in a peer-reviewed journal was included, and a few dissertations from university publications. These peer-reviewed journals were indexed in PsychINFO, Sage, SCOPUS, Taylor & Francis databases. These researches were then studied, to figure out a pattern of spiraling in WI, and the impact of COVID work culture on the same.

3.3 THEORETICAL FRAMEWORK

Through the studies that have tried to explain workplace incivility, a rich text for theorizing the same has been developed. The impact of organizational and dispositional factors in instigating and experiencing incivility has been studied vehemently. Various antecedents have been identified, which directly and indirectly impact workplace incivility.

Antecedents for uncivil behavior in a workplace are categorized on the basis of their origin source: organizational, situational, and individual.

Organizational antecedents have been studied well over the past two decades, and their arch has slowly moved from informal attitude (Andersson & Pearson, 1999; Muir, 2000; Gardner & Johnson, 2001) to technology (Muir, 2000; Vickers, 2006). Informal attitude or work norms that don't specify formal boundaries give enough lee way to make unassessed behavioral errors that could be deemed as uncivil by workers. Being available all the time or having access to co-workers and the work environment through e-mails and video conferencing puts psychological strain on an individual and is theorized to be liable for uncivil work climate. Other organizational antecedents concluded upon include hierarchical structure (Muir, 2000), stagnant wages (Gardner & Johnson, 2001), contract violation (Blau, 2007; Sears & Humiston, 2015), work exhaustion (Blau & Andersson, 2005), and job insecurity (Salin, 2003; Torkelson et al., 2016). Downsizing (Andersson & Pearson, 1999; Gardner & Johnson, 2001; Buhler, 2003) will evidently lead to increase in work pressure, job demands (Van Jaarsveld et al., 2010; Torkelson et al., 2016; Koon & Pun, 2018), and job insecurity. Organization change and globalization (Gardner

& Johnson, 2001; Muir, 2000) and globalization will also agitate the workers, as both of these signify a change in the work climate. Poor social support from workers (Holm et al., 2015; Torkelson et al., 2016) can turn out to be the biggest factor in turnover or workplace incivility incidents. During COVID pandemic, the impact of being present at work while being sick resulted in low incivility experience than usual out of concern for the co-worker's health (Taylor et al., 2021). The same study also found that with an increase in workload, co-worker presenteeism leads to increased incivility out of self-concern.

These factors clearly indicate that organizational antecedents won't be the sole reason for workplace incivility. Some personality types and dispositions are more susceptible to incidents of incivility due to perceptions and cultural precedents or prejudices.

Personality factors or temperament of impulsivity, Reactivity/high sensitivity and rebellious nature have been creed as more susceptible to an uncivil interaction in the workplace (Andersson & Pearson, 1999). Further, studies depicting aggressive personalities and high functioning individuals such as that of Type A personality experience more uncivil interactions in the workplace (Salin, 2003). Studies also depict those individuals with low agreeableness experience high incivility, and those who are neurotic have high perceived incivility (Milam et al., 2009). Poor reciprocity and narcissism are also positively related to workplace incivility (Meier & Semmer, 2013).

Research also shows that individuals with workaholic tendencies, and low psychological capital tend to engage more in Workplace incivility (Lanzo et al., 2016). These individuals also tend to perceive higher incivility at workplace, which shows that an informal work culture around them could increase the chances of negative workplace interactions.

Narcissistic individuals respond aggressively to incivility (Chen et al., 2013; Pei et al., 2019) and hence instigate workplace incivility. Individuals with high social anxiety experience more uncivil behavior (Brown, 2021). Coming to gender disposition, research have shown that women experience more incivility and are more likely be targeted with incivility (Cortina et al., 2001, 2013; Milam et al., 2009; Young et al., 2019). Race has also been studied and research (Cortina et al., 2013) shows that minority race groups tend to experience higher incivility.

Thus, individuals with high narcissistic traits and Type A personality will most likely be the instigator of workplace incivility, targeting individuals with high agreeableness, women, and minority race groups as they seem the weaker links in any group.

The organizational and individual factors which lead to or aid the incidents of workplace incivility are often times given ripe conditions such as counterproductive work behavior (Meier & Spector, 2013), role ambiguity (Taylor & Kluemper, 2012) and leadership (Ghosh et al., 2011; Lee & Jensen, 2014; Young et al., 2019). These are known as situational factors or situational antecedents. These situational factors act as mediators when it comes to incidents of workplace incivility (Figure 3.1).

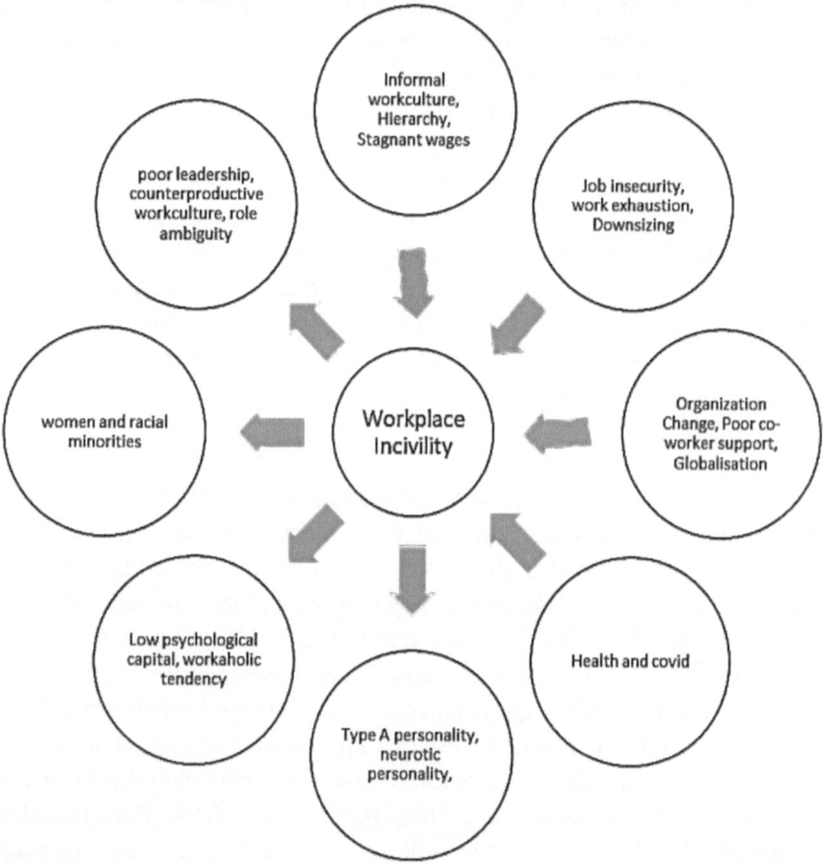

FIGURE 3.1 Image depicts the various factors which mediate and provide habitable conditions for incidents of workplace incivility.

Studying what causes or leads to incidents of workplace incivility leaves room open for studying the consequences of the same. It is interesting to see how through the consequences of workplace incivility, the other factors come full circle.

3.4 OUTCOMES

As supported by studies above, experiencing, and witnessing workplace incivility leads to more incidents of incivility, but they also lead to consequences of professional and personal criteria.

3.4.1 ORGANIZATIONAL OUTCOMES

The most commonly noted outcome of incidents of workplace incivility is turnover intentions (Andersson & Pearson, 1999; Muir, 2000; Gardner & Johnson, 2001; Pearson & Porath, 2005; Salin, 2003; Lim et al., 2008; Lim & Teo, 2009; Cortina et al., 2013; Rahim & Cosby, 2016; Sharma & Singh, 2016; Taylor et al., 2017), loss of profit directly (Gardner & Johnson, 2001) or indirectly (Porath & Pearson, 2013). This indirect loss usually happens due to poor performance (Pearson et al., 2000; Hutton & Gates, 2008; Chen et al., 2013; Gedro & Wang, 2013; Rahim & Cosby, 2016; Schilpzand & Huang, 2018). Research (Shin et al., 2021) shows that the impact of customer incivility increased on job performance in post-pandemic work culture. Workplaces with poor job benefits and less growth also experience higher turnover intentions. The impact of workplace incivility on creating a negative work environment is studied in Ghosh et al. (2011). With negative social interactions such as workplace incivility or bullying, the negative affect increases and the witnesses also become a part of the incident. The decline in positive affect corrodes the work atmosphere, increasing the likelihood of more negative interactions and low job performance. The organizational loss can be summed up in terms of loss of workforce and hence, the loss of capital.

3.4.2 INDIVIDUAL OUTCOMES

The impact of workplace incivility on physical health (Cortina et al., 2001; Lim et al., 2008; Miner et al., 2012), mental health (Lim et al., 2008; Miner et al., 2012) has been studied quite extensively over the past two decades. This research shows that workplace incivility not only disturbs the mental peace of an individual but also affect their physical health. Job strain increases life stress and anxiety in individuals. Mental health problems usually manifest in the form of poor social relations and physical symptoms of heart issues and fluctuating energy levels.

The more refined aspects of an individual's life such as job satisfaction (Blau & Andersson, 2005; Lim & Teo, 2009; Miner-Rubino & Reed, 2010; Holm et al., 2019), job stress (Miner et al., 2012; Koon & Pun, 2018), and life satisfaction (Miner et al., 2012) have also been shown to be impacted negatively by Workplace Incivility. With the increase in job stress, and consequent decrease in job and life satisfaction, the decline in one's health and turnover intentions will increase.

The more novel concept of burnout has also been researched in reference to Workplace incivility (Miner-Rubino & Reed, 2010; Taylor et al., 2017), along with emotional exhaustion (Kern & Grandey, 2009). Staying in an unproductive and unhealthy work environment for long will lead to emotional exhaustion and subsequent burnout. This will indeed affect one's personal life, their marital life affairs, and other interpersonal connections. Finally, the impact of workplace incivility on family support, family relations, marital relations and satisfaction has also been researched vehemently (Ferguson, 2012; Pei et al., 2019) in the past decade, showcasing a damaging effect. Ferguson (2012) showed that people felt that their partners after experiencing incivility at the workplace were displacing their feelings at home.

Being disturbed due to negative interactions at the workplace would lead to slow and sure deterioration of one's overall health, and hence, chances of engaging in negative interactions will keep on rising. An individual who is susceptible to negative perceptions, or more likely to be targeted for uncivil conversations will end up instigating or inspiring workplace incivility. Research shows that experiencing Workplace Incivility or Witnessing workplace incivility has also been deemed as an effective antecedent for incidents of incivility in the workplace (Andersson & Pearson, 1999; Cortina et al., 2001; Blau & Andersson, 2005; Lim et al., 2008; Trudel & Reio, 2011; Holm et al., 2015; Torkelson et al., 2016). And when this is combined with certain personality traits such as narcissism and workaholic culture, chances of instigating incivility increase (Figure 3.2).

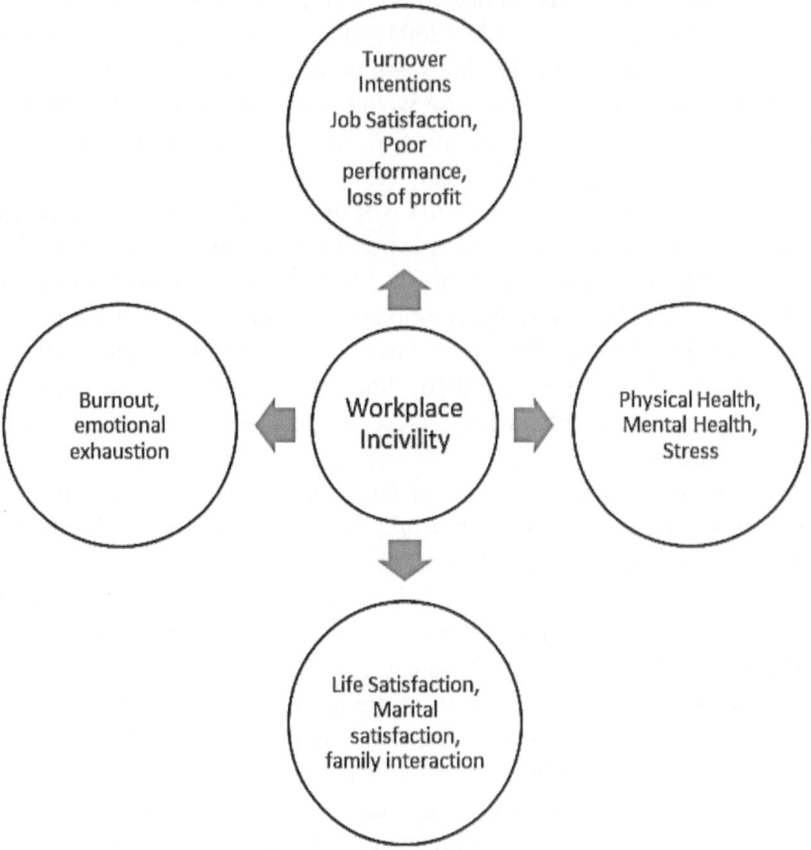

FIGURE 3.2 Image depicts the various aspects of an individual's life that are impacted by workplace incivility.

3.4.3 TECHNOLOGY AND WORKPLACE INCIVILITY

With the current COVID-19 pandemic, the influence of technological advancements in the organization sector cannot be overlooked. With a completely new platform to communicate on, the behavioral aspects and norms also get modified and require new markers for uncivil behavior. To add to the stress of the job, the healthcare stress also surmounted the lives of workers these two years, and hence, it's important to note how the online communication channel are affecting one's health and lifestyle.

Park et al. (2015) has studied Cyber workplace incivility as a growing stressor in the work place. Suler (2004) talks about an online disinhibition effect showcasing how individuals end up doing and saying things on a cyber platform that they otherwise would never have engaged in. Due to lack of physical cues, the perception of these messages being communicated via online platforms also changes.

Cyber incivility through e-mails will include discourteous messages, using e-mails to give short-notices, expecting a 24×7 open communication line and the same. Giumetti et al. (2013) reveal the negative impact of cyber workplace incivility on emotional exhaustion and positive affect. Cyberincivility has also been linked to poor job satisfaction and poor job performance (Giumetti et al., 2016). Staying available to work life from home can be detrimental when looked at the lack of peer support and social engagement.

The studies show a clear set of antecedents, outcomes, and putting technology as a mediator for workplace behaviors, the relationship between these factors establishes itself.

With workplace incivility having individual outcomes such as burnout, job dissatisfaction, poor performance, family disruptions, and low life satisfaction, the individual is bound to feel irritable, exhausted, and anxious. Based on the personality of the individual, these circumstances which are outcomes of WI, would serve as perfect antecedents for the primary spiral of workplace incivility incidents.

During COVID, when the workplace went virtual, the employment of technology in determining workplace culture was also quite significant. The studies have shown that virtual workplace demands a different and much more strict decorum, due to its option of 24×7 availability.

3.4.4 METHODS

Methods used to study workplace incivility have mostly been empirical, relying on measures such as organizational tolerance workplace incivility (OTWI) (Hulin et al., 1996), modified version of OTWI (Loi et al., 2015), workplace incivility scale (Cortina et al., 2001), enacted incivility scale (Cortina et al., 2013), nursing incivility scale (NIS) (Guidroz et al., 2007), norms for civility scale (Walsh et al., 2012), cyber incivility scale (Lim et al., 2009), and gendered incivility scale (Miner et al., 2012). Workplace Incivility Scale has been adapted numerous times to meet the needs of specific research, with the latest adaptation utilized in Samma et al. (2020). These researches have involved specific domain sample to general workplace samples, to figure out the relationship of WI with its various facets. However, to study interventions, the preferred mode has been experimental.

Some subjective methods have also been employed to study Workplace incivility. These range from interviews (Cortina et al., 2001), reacting to past incidents of workplace incivility or reacting to hypothetical incidents of incivility (Porath & Pearson, 2012). Other researchers have utilized a modified version of the same, such as deciphering hypothetical workplace incivility incidents or formulating empirical responses for similar hypothetical situations.

3.5 FINDINGS

Author's Name	Year	Journal Name	Sample	Theoretical Function of WI	Major Findings
Miner-Rubino & Reed	2010	Journal of Applied Social Psychology	Property management employees	Antecedent	Workplace incivility led to turnover intentions, burnout, and job satisfaction.
Van Jaarsveld et al.	2010	Journal of Management	Call Center Employees	Antecedent	Experienced customer incivility was positively correlated with higher job demands and emotional exhaustion, which led to higher instigated WI towards customers.
Ghosh et al.	2011	Advances in Developing Human Resources	Different sector employees	Consequence	The perception of mentor's distancing behavior and the frequency of instigated incivility in protégé's behavior against mentors was mediated by negative affect.
Trudel & Reio	2011	Human Resource Development Quarterly	Manufacturing and health care employees	Outcome	Instigated incivility is positively related to being an incivility target.
Ferguson	2012	Journal of Organizational Behavior	Employees and their partners	Antecedent	Workplace Incivility led to strained relationships with their partners, impacting the partner's work life as well.
Miner et al.	2012	Journal of Applied Social Psychology	Management employees and undergraduate students	Antecedent	Organization support and emotional support helped deal with the experience of workplace incivility.

Author's Name	Year	Journal Name	Sample	Theoretical Function of WI	Major Findings
Chen et al.	2013	Academy of Management Journal	Full-time employees	Antecedent	Narcissism moderated the effect of incivility on engagement.
Cortina et al.	2013	Journal of Management	Law enforcement, municipality, and military employees	Antecedent	Experienced incivility is associated with turnover intentions.
Gedro & Wang	2013	Advances in Developing Human Resources	Employees	Antecedent	Development of a training program.
Giumetti et al.	2013	Journal of Occupational Health Psychology	Undergraduate students	Antecedent	Cyber incivility resulted in lower positive affect, task performance and engagement, and a higher negative affect.
Meier & Summer	2013	European Journal of Work and Organizational Psychology	Employees	Consequence	Lack of reciprocity was positively related to anger which was positively related to WI.
Meier & Spector	2013	Journal of Applied Psychology	Employees	Consequence/ antecedent	Counterproductive work behavior has a reciprocal relationship with work stressors.
Porath & Pearson	2013	Harvard Business Review	Managers and employees	Antecedent	Incivility is expensive, and only a few organizations take action to recognize and curb it.
Lee & Jensen	2014	Group and Organization Management	Employee-coworker dyads	Consequence	Active constructive leadership decreased evidence of workplace incivility.
Reich & Hershcovis	2014	Journal of Applied Psychology	Employees	Antecedent	Witnesses of WI tend to punish instigators.
Holm et al.	2015	BioMed Research International	Employees	Consequence	Workplace incivility was associated with instigated incivility and negative outcomes.
Park et al.	2015	Journal of Management	Employees	Antecedent	Experiencing cyber incivility caused higher affective and physical distress in workers, impacting the next working day. Job control and psychological detachment from work aided in reducing work distress.

Author's Name	Year	Journal Name	Sample	Theoretical Function of WI	Major Findings
Sears & Humiston	2015	Journal of Managerial Psychology	Employees	Consequence	Psychological contract violation resulted in higher levels of uncivil workplace behavior.
Giumetti et al.	2016	Psi Chi Journal of Psychology	Employees	Antecedent	Cyber incivility was negatively correlated to job performance and Job satisfaction. Positive affect acted as a buffer against the impact of cyber incivility on job performance.
Lanzo et al.	2016	International Journal of Workplace Health Management	Employees	Consequence	Workaholics reported higher levels of stress, low levels of psychological capital, and consequently instigated workplace incivility.
Rahim & Cosby	2016	The Journal of Management Development	Employees	Antecedent	WI is negatively associated with job performance (negatively), job burnout (positively) and turnover intentions (positively).
Schilpzand et al.	2016	Journal of Organizational Behavior	Undergraduate and business students	Antecedent	Sharing the experience of WI with another target reduced the negative consequences.
Sharma & Singh	2016	South Asian Journal of Global Business Research	Restaurant employees	Antecedent	Experience of Wi resulted in interpersonal distress, lower job satisfaction.
Torkelson et al.	2016	Work & Stress	Employees	Consequence	Perpetration of WI had a direct link with organization variables – job insecurity, low social support from co-workers, job demands, organizational change.
Taylor et al.	2017	Journal of Management	Non-profit organization employees	Antecedent	Incivility changes depicted variation in burnout, emotional exhaustion, and turnover intentions.
Koon & Pun	2018	The Journal of Applied Behavioral Science	University academic staff	Consequence	Higher job demands resulted in higher emotional exhaustion, leading to lower job satisfaction, consequently resulting in the instigation of workplace incivility.

Author's Name	Year	Journal Name	Sample	Theoretical Function of WI	Major Findings
Schilpzand & Huang	2018	Journal of Applied Psychology	Employees and their supervising managers	Antecedent	Employees identifying with their workplace experienced higher negative affect when experiencing ostracization at the workplace.
Anju & Raaj	2019	International Journal of Engineering and Advanced Technology	Hospital nurses	Antecedent	Perception of WI has significant effects of work-related outcomes like Emotional exhaustion and Turnover intentions.
Holm et al.	2019	International Journal of Workplace Health Management	Trade union employees	Antecedent	Witnessed workplace incivility results in instigated workplace incivility. Witnessing WI is associated with both perceived stress and low job satisfaction. Perceiving high levels of control and social support from co-workers resulted in individuals expressing higher levels of instigated incivility on witnessing co-worker incivility.
Pei et al.	2019	Personality and Individual Differences	Employees	Antecedent	Individuals who recalled incidents of workplace incivility reported lower levels of family satisfaction. Narcissistic individuals were highly reactive to incidents of WI, and consequently experienced lower levels of family satisfaction.
Samosh	2019	Organizational Management Journal	Participants	Consequence	Examples of WI include criticizing, dismissing, insulting, shouting, inappropriate language, indirect talk, name-calling, unfair treatment, inappropriate jokes, and fighting. Arguing, lying, invading privacy, backstabbing, intimidating, threatening, and withholding information.
Young et al.	2019	Public Management Review	Government and non-profit employees	Consequence	Women are more likely to experience WI as compared to men, but this is attenuated with ethical leadership. Personal control and ethical leadership are negatively associated with experiences of WI.

Author's Name	Year	Journal Name	Sample	Theoretical Function of WI	Major Findings
Hulsheger et al.	2020	Journal of Applied Psychology	Employees	Consequence	High mindfulness resulted in low and less variant incivility. High in train mindfulness individuals also reacted morally to events of instigated incivility and experienced guilt, as compared to individuals low in trait mindfulness.
Kanitha & Naik	2020	International Journal of Nursing Education	Staff nurses	Consequence	82% of staff nurses had adequate knowledge of WI, and 77% of nurses experienced workplace Incivility.
Megeirhi et al.	2020	Journal of Hospitality and Tourism Management	Hotel employees	Antecedent	Employees perceiving high levels of organizational tolerance for incivility increased job search behavior.
Samma et al.	2020	Healthcare	Workers of small and medium-sized entrepreneurs	Antecedent	Workplace incivility has a negative impact on innovative work behavior.
Brass	2021	Studies by undergraduate researchers at Guelph	Employees	Antecedent	The results showed that although incivility does lead to job burnout and workplace anxiety in virtual, the prediction that these experiences would be more frequent was not supported.
Brown	2021	Electronic Depository and Dissertation Repository	Employees	Consequence	Individuals with high social anxiety reported more incidents of WI. However, participants reported more positive emotions when they were socially anxious. Self-efficacy added to this positive effect.
Greenslade	2021	University of New Zealand	Employees	Antecedent	Secondary spirals are more common in the workplace. Supervisor's actions are the primary source of incivility spirals.

Author's Name	Year	Journal Name	Sample	Theoretical Function of WI	Major Findings
Shin et al.	2021	Service Business	COVID frontline service employees	Antecedent	Customer incivility indirectly impacted job performance and directly impacted emotional exhaustion post-pandemic. Emotional exhaustion was a smoother mediator for the impacts of customer incivility than abusive supervision.
Taylor et al.	2021	Journal of Applied Psychology	Service employees	Consequence	Co-worker presenteeism decreased workplace incivility out of concern if the individual was sick. But with higher workloads, this result reversed, and co-worker mistreatment increased.

3.6 DISCUSSION

With the beginning of research in the topic of workplace incivility, Andersson & Pearson (1999) also suggested a sort of spiral originating from the inter-relation of the variables surrounding the phenomenon of WI. They proposed that individuals with high reactivity will respond negatively to slight misconceptions and might be involved in more negative interactions at the workplace. For environmental factor, they considered uncomfortable climate, and individuals who had once experienced any incident of WI would be more likely to repeat the incident in the same environment. With everyone involved reacting and creating new incidents of incivility in the workplace, a spiraling effect was showcased.

This spiraling aspect of WI has been noted and studied for some time, based on the antecedents and outcomes. However, with COVID and technological advancements, a completely new situation has emerged. To understand the spiraling effect of WI in the current times, it is crucial that one considers both the established antecedents/outcomes and the novel variables. One of these novel variables has been an unending, constant health anxiety, and grief for the unsurmountable loss of life that has occurred in the past two years. Along with this, the cutting-off of social ties, and the workplace going completely online has not made things easier. Many organizations went through huge losses, turnovers, and were shut down.

This has only increased psychological stress manifold. Adding to this mix, is the technological aid that made virtual work-life a constant source of distress. It is possible to monitor the employees while they work, micro-manage them, and expect them to be available to work at all hours. These factors not only make the individuals irritable, highly reactive, and distressed, but also lower the morale of the work force.

Past few studies have shown that during COVID, the workforce has been more reactive towards negative incidents, including customer and supervisor incivility. This would only increase turnover intentions on the part of the work-force.

The studies showed one important aspect of considering the impact of WI incidents on not only the workers, but also on the organization as a whole, so as to get the management to inculcate programs of interventions.

Studying these factors shows how the different aspects of WI form a web of sorts, with each variable impacting the next.

Family disruptions/marital conflict ⟹ *Emotional reactivity, stress* ⟹ *Emotional exhaustion/irritability* ⟹ *Workplace incivility* ⟹ *Psychological stress* ⟹ *Emotional exhaustion* ⟹ *Job burnout/low satisfaction* ⟹ *Family disruptions/marital conflict*

Organizational change ⟹ *Increase in job demands/stagnant wages* ⟹ *Job distress* ⟹ *Workplace incivility* ⟹ *Job burnout* ⟹ *Turn-over intentions* ⟹ *Loss of profit* ⟹ *Organizational change*

These two instances showcase the loop of workplace incivility, with its outcomes finally becoming its antecedents again.

Considering the impact of the pandemic, the organizational change, grief, psychological distress, job demands, stagnant wages, health complications, financial distress are some of the constants which provide an adequate environment for incidents of incivility to pan out in any workplace. With such constants present, one step in these spirals is reduced, which in turn makes it easier for these spirals to occur in a frame.

The study proposes that pandemic provides adequate conditions to facilitate the incidents of WI, and hence, can be considered as an antecedent in itself for Workplace Incivility.

3.7 CONCLUSION

The research was conducted to encompass the concept of workplace incivility and its hypothesized frequency in pandemic. The literature was reviewed from both the pre-COVID and post-COVID era to consider the role COVID conditions would play in the phenomenon of WI. The literature was accessed through various databases, taking into account only peer-reviewed research. Pre-COVID data was richly filtered to avoid any studies with replicating and similar results. Established variables affecting the instances and influences of WI were presented through single studies to reduce to litter. A map of the chaining relationship between the antecedents and outcomes of Workplace incivility was created, to understand the spiraling nature they posited. Considering the variables which were enhanced in frequency and intensity under the pandemic, the chapter proposed that the spiraling of the incidents would inherently lose one or more steps in the process as the pandemic conditions provided the adequate environmental and situational antecedents for incidents of WI.

3.8 LIMITATIONS

The study was a literature review and hence, did not empirically test the proposed impact of COVID on WI spiral.

The study could not entail much data post-pandemic, as there was a limited research bank available for this period.

The study also did not include the interventions applied and their role, as post-pandemic intervention data was rare.

3.9 FUTURE RECOMMENDATIONS

The studies only referred to workplace incivility, but considering the wide net this variable has cast, some significant antecedents/outcomes should be taken as secondary variables.

There wasn't enough literature present in post-COVID time period outside of the healthcare sector. Other areas, such as education and finance, should also have been vehemently researched.

The research presents COVID conditions as adequate antecedents for WI. This proposal should be empirically tested.

KEYWORDS

- **antecedents**
- **globalization and capitalization**
- **meta-analysis**
- **methodology**
- **nursing incivility scale**
- **outcomes**
- **post-Covid**
- **work culture**
- **workplace incivility**

REFERENCES

Andersson, L. M., & Pearson, C. M., (1999). Tit for tat? The spiraling effect of incivility in the workplace. *The Academy of Management Review, 24*(3), 452–471.

Anju, E. N., & Raaj, V. G., (2019). Unseen incivility in the workplace and its impact on work allied outcomes. *International Journal of Engineering and Advanced Technology*, 80–85.

Blau, G., & Andersson, L. M., (2005). Testing a measure if instigated workplace incivility. *Journal of Occupational and Organizational Psychology*, 595–614.

Blau, G., (2007). Partially testing a process model for understanding victim responses to an anticipated worksite closure. *Journal of Vocational Behavior*, 401–428.

Brass, J., (2021). Incivility in relationship formation during virtual work. *Studies by Undergraduate Researchers at Guelph.*

Brown, J., (2021). More than just shying away from conflict: The relationship between social anxiety and workplace anxiety. *Electronic Depository and Dissertation Repository*. The University of Western Ontario.

Buhler, P., (2003). Managing in the new millennium. *Supervision*, 20.

Chen, Y., Ferris, D. L., Kwan, H. K., Yan, M., Zhou, M., & Hong, Y., (2013). Self-love's lost labor: A self-enhancement model of workplace incivility. *Academy of Management Journal*, 1199–1219.

Cortina, L. M., Kabat-Farr, D., Leskinen, E. A., Huerta, M., & Magley, V. J., (2013). Selective incivility as modern discrimination in organizations evidence and impact. *Journal of Management*, 1570–1605.

Cortina, L. M., Magley, V. J., Williams, J. H., & Langout, R. D., (2001). Incivility in the workplace: Incidence and impact. *Journal of Occupational Health Psychology*, 64–80.

Estes, B., & Wang, G., (2008). Workplace incivility: Impacts on individual and organizational impacts. *Human Resource Development Review*.

Felblinger, D. M., (2008). Incivility and bullying in the workplace and nurses' shame responses. *Journal of Obstetric, Gynaecological, and Neonatal Nursing*, 234–242.

Ferguson, M., (2012). You cannot leave it at the office: Spillover and crossover of co-worker incivility. *Journal of Organizational Behavior*, 571–588.

Gardner, S., & Johnson, P. R., (2001). The leaner, meaner workplace: Strategies for handling bullies at work. *Employment Relations Today*, 269–286.

Gedro, J., & Wang, J., (2013). Creating civil and respectful organizations through the scholar-practitioner bridge. *Advances in Developing Human Resources*, 284–295.

Ghosh, E., Derkes, S., & Falletta, S., (2011). Incivility spiral in mentoring relationships, reconceptualizing negative mentoring as deviant workplace behavior. *Advances in Developing Human Resources*, 22–39.

Giumetti, G., Muth, E. R., & Schroeder, A., (2013). What a rude e-mail! Examining the differential effects of incivility versus support on mood, energy, engagement and performance in an online context. *Journal of Occupational Health Psychology*, 297–309.

Giumetti, G., Saunders, L. A., Brunette, J. P., DiFrancesca, F. M., & Graham, P. G., (2016). Linking cyber incivility with job performance through job satisfaction: The buffering role of positive affect. *Psi Chi Journal of Psychology*, 230–240.

Greenslade, J., (2021). *The Spiraling Effects of Witnessing Incivility in the Workplace (Master's Thesis)*. New Zealand: University of Canterbury.

Guidroz, A. M., Burnfield, J. L., Clark, O. I., Schwetschenau, H. M., & Jex, S., (2007). *The Nursing Incivility Scale: Development of an Occupation-Specific Incivility Measure*. Poster presented at the 22nd annual conference of the society for industrial and organizational psychology. New York, NY.

Holm, K., Torkelson, E., & Backstorm, M., (2015). Models of workplace incivility: The relationships to instigated incivility and negative outcomes. *BioMed Research International*, 10.

Holm, K., Torkleson, E., & Backstron, M., (2019). Exploring links between witnessed and instigated workplace incivility. *International Journal of Workplace Health Management*, 160–175.

Hulin, C. L., Fitzgerald, L. F., & Drasgow, F., (1996). Organizational influences on sexual harassment. In: Stockade, M. S., (ed.), *Sexual Harassment in the Workplace: PERSPECTIVES, Frontiers, and Response Strategies* (pp. 127–150). Thousand Oaks, CA: Sage.

Hülsheger, U. R., Gils, S. V., & Waikowiak, A., (2020). The regulating role of mindfulness in enacted workplace incivility: An experience sampling study. *Journal of Applied Psychology*.

Hutton, S., & Gates, D., (2008). Workplace incivility and productivity losses among direct care staff. *Workplace Health and Safety*, 168–175.

Kanitha, D., & Naik, P. R., (2020). Knowledge and frequency of experience of workplace incivility among the staff nurses. *International Journal of Nursing Education*, 48–53.

Kern, J. H., & Grandey, A. A., (2009). Customer incivility as a social stressor: The role of race and racial identity for service employees. *Journal of Occupational Health Psychology*, 46–57.

Koon, V. Y., & Pun, P. Y., (2018). The mediating role of emotional exhaustion and job satisfaction on the relationship between job demands and instigated workplace incivility. *The Journal of Applied Behavioral Science*, 187–207.

Lanzo, L., Aziz, S., & Wuensch, K., (2019). Workaholism and incivility: Stress and psychological capital's role. *International Journal of Workplace Health Management,* 165–183.

Lee, J., & Jensen, J. M., (2014). The effects of active constructive and passive corrective leadership on workplace incivility and the mediating role fairness perceptions. *Group and Organization Management*, 416–443.

Lewis, P. S., & Malecha, A., (2011). The impact of workplace incivility on the work environment, manager skill, and productivity. *The Journal of Nursing Administration, 41*(1), 41–47.

Lim, S., Cortina, L. M., & Magley, V. J., (2008). Personal and workgroup incivility: Impact on work and health outcomes. *Journal of Applied Psychology*, 95–107.

Lim, V. K., & Teo, T. S., (2009). Mind your e-manners: Impact of cyber incivility on employees' work attitude and behavior. *Information & Management,* 419–425.

Loi, N. M., Loh, J. M., & Hine, D. W., (2015). Don't rock the boat: The moderating role of gender in the relationship between workplace incivility and work withdrawal. *The Journal of Management Development*, 169–186.

Megeirhi, H. A., Ribeiro, M. A., & Woosnam, K. M., (2020). Job search behavior explained through perceived tolerance for workplace incivility, cynicism and income level: A moderated mediation model. *Journal of Hospitality and Tourism Management*, 88–97.

Meier, L. L., & Semmer, N. K., (2013). Lack of reciprocity, narcissism, anger, and instigated workplace incivility: A moderated mediation model. *European Journal of Work and Organizational Psychology*, 461–475.

Meier, L. L., & Spector, P., (2013). Reciprocal effects of work stressors and counterproductive work behavior: A five-wave longitudinal study. *Journal of Applied Psychology,* 529–539.

Milam, A. C., Spitzmueller, C., & Penney, L. M., (2009). Investigating individual differences among targets of workplace incivility. *Journal of Occupational Health Psychology, 14*(1), 58–69.

Miner, K. N., Settles, I. H., Pratt-Hyatt, J. S., & Brady, C. C., (2012). Experiencing incivility in organizations: The buffering effects of emotional and organizational support. *Journal of Applied Social Psychology*, 340–372.

Miner-Rubino, K., & Reed, W. D., (2010). Testing a moderated mediational model of workgroup incivility: The roles of organizational trust and group regard. *Journal of Applied Social Psychology*, 3148–3168.

Muir, C., (2000). Can we all get along? The interpersonal challenge at work. *Academy of Management Executive*, 143, 144.

Park, Y., Fritz, C., & Jex, S. M., (2015). Daily cyber incivility and distress: The moderating role of resources at work and home. *Journal of Management,* 1–23.

Pearson, C. M., & Porath, C. L., (2005). On the nature, consequences and remedies of workplace incivility: No time for "nice"? Think again. *The Academy of Management Executive, 19*(1), 7–18.

Pearson, C. M., Andersson, L. M., & Porath, C. L., (2000). Assessing and attacking workplace incivility. *Organizational Dynamics*, 123–137.

Pearson, C. M., Andersson, L. M., & Porath, C. L., (2005). Workplace incivility. In: Fox, S. A., (ed.), *Counterproductive Workplace Behavior: Investigations of Actors and Targets* (pp. 177–200). Washington, DC: American Psychological Association Press.

Pearson, C., Andersson, L., & Wegner, J., (2001). When workers flout convention: A study of workplace incivility. *Human Relations*, 1387–1419.

Pei, L., Chenjie, X., Jinglun, H., Xiaotian, W., & Aimei, L., (2019). Experienced workplace incivility, anger, guilt, and family satisfaction: The double-edged effect of narcissism. *Personality and Individual Differences*.

Porath, C. L., & Pearson, C. M., (2012). Emotional and behavioral responses to workplace incivility and the impact of hierarchical status. *Journal of Applied Social Psychology*, 326–357.

Porath, C., & Pearson, P., (2013). The price of incivility. *Harvard Business Review*, 115–121.

Rahim, A., & Cosby, D. M., (2016). A model of workplace incivility, job burnout, turnover intentions and job performance. *The Journal of Management Development*, 1256–1265.

Reich, T. C., & Hershcovis, M. S., (2014). Observing workplace incivility. *Journal of Applied Psychology*.

Ricciotti, N. A., (2016). Emotional intelligence and instigation of workplace incivility in a business organization. *Walden Dissertations and Doctoral Studies*.

Salin, D., (2003). Ways of explaining workplace bullying: A review of enabling, motivating and participating structures and processes in the work environment. *Human Relations*, 1213–1232.

Samma, M., Zhao, Y., Rasool, S. F., Han, X., & Ali, S., (2020). Exploring the relationship between innovative work behavior, job anxiety, workplace ostracism, and workplace incivility: Empirical evidence from small and medium sized enterprises. *Healthcare*, 1–15.

Samosh, J., (2019). What is workplace incivility? An investigation of employee relational schemas. *Organizational Management Journal*, 81–97.

Schilpzand, P., & Huang, L., (2018). When and how experienced incivility dissuades proactive performance: An integration of sociometer and self-identity orientation perspectives. *Journal of Applied Psychology*, 828–841.

Schilpzand, P., De Pater, I. E., & Erez, A., (2016). Workplace incivility: A review of the literature and future agenda research. *Journal of Organizational Behavior*, 57–88.

Sears, K., & Humiston, G. S., (2015). The role of emotion in workplace incivility. *Journal of Managerial Psychology*, 390–405.

Sharma, N., & Singh, V. K., (2016). Effects of workplace incivility on job satisfaction and turnover intentions in India. *South Asian Journal of Global Business Research, 5*(2).

Shin, Y., Hur, W. M., & Hwang, H., (2021). Impacts of customer incivility and abusive supervision on employee performance: A comparative study of the pre-and post-COVID-19 periods. *Service Business*.

Suler, J., (2004). The online disinhibition effect. *Cyberpsychology & Behavior*, 321–326.

Taylor, S. G., & Kluemper, D. H., (2012). Linking perceptions of role stress and incivility to workplace aggression: The moderating role of personality. *Journal of Occupational Health Psychology*, 316–329.

Taylor, S. G., Bedeian, A. G., Cole, M. S., & Zhang, Z., (2017). Developing and testing a dynamic model of workplace incivility change. *Journal of Management*, 645–670.

Taylor, S. G., Butts, M. M., Cole, M. S., & Pounds, T. W., (2021). Are you sick? Understanding the effects of co-worker presenteeism on workplace mistreatment. *Journal of Applied Psychology*, 1299–1313.

Torkelson, E., Holm, K., Backstrom, M., & Schad, E., (2016). Factors contributing to the perpetration of workplace incivility: The importance of organizational aspects and experiencing incivility from others. *Work & Stress*, 115–131.

Trudel, J., & Reio, T. G. Jr., (2011). Managing workplace incivility: The role of conflict management styles-antecedents or antidote? *Human Resource Development Quarterly*, 395–423.

Van, J. D. D., Walker, D. D., & Skarlicki, D. P., (2010). The role of job demands and emotional exhaustion in the relationship between customer and employee incivility. *Journal of Management*, 1486–1504.

Vasconsales, A. F., (2020). Workplace incivility: A literature review. *International Journal of Workplace Health Management*.

Vickers, M. H., (2006). Writing what's relevant: Workplace incivility in public administration – A wolf in sheep's clothing. *Administrative Theory & Praxis*, 69–88.

Walsh, B. M., Magley, V. J., Reeves, D. W., Davies-Schrils, K. A., Marmet, M. D., & Gallus, J. A., (2012). Assessing workgroup norms for civility: The development of the civility norms questionnaire-brief. *Journal of Business and Psychology*, 407–420.

Young, K. A., Hassan, S., & Hatmaker, D. M., (2019). Towards understanding workplace incivility: Gender, ethical leadership and personal control. *Public Management Review*.

CHAPTER 4

Changes of Collaboration Platforms' Use Perceived by Users Because of COVID: A Research Study of Changes in User Experience of Microsoft Teams Platform

ZDENKO DEŽELAK, SIMONA STERNAD ZABUKOVŠEK, and SAMO BOBEK

University of Maribor, Faculty of Economics and Business, Slovenia

ABSTRACT

Collaboration platforms have become very important in recent years. During the COVID pandemic, they became crucial for many organizations while they are infrastructure for remote work. Because of this, employees started to use collaboration platforms on different, usually more advanced level. This affects employees' attitudes toward collaboration platforms and how employees perceive new technology used as collaboration platforms users. Acceptance of collaboration platforms can be researched using user experience (UX) concepts. UX covers all aspects of digital products and services that users feel sense directly while getting to know them and regularly using them. In our research, we have chosen the UEQ questionnaire, which measures the impression of user experience. Standard usability aspects, such as dependability, efficiency, and perspicuity, as well as user experience aspects such as originality and stimulation, are assessed. The collected data shows that the COVID pandemic has impacted the UX of

Reshaping the Business World Post-COVID-19: Management Strategies for Sustainable Behavior Change. Arvind K. Birdie & Ruchi Joshi, (Eds.)

Microsoft Teams users researched. Because of the fact that more functionality was being used compared to the pre-COVID period, the UX values of the measured scales have increased, so the overall UX has gotten better. This can be confirmed with the Attractiveness scale, which significantly bumped up compared to pre-COVID data. Also, pragmatic quality has increased. on the contrary, hedonic quality has slightly decreased.

4.1 INTRODUCTION

Collaboration platforms have become very important in recent years. Even before the COVID pandemic, new ways of collaboration have emerged, facilitated by the rapid advances in information technology innovations. New IT solutions and applications are being placed on the market regularly, and organizations are always seeking new sources of competitive advantage, more efficient ways of doing business, and communicating. COVID pandemic has made all this even more obvious and important. Organizations had to switch to remote work almost overnight. Although many of them have already used remote forms of work to some extent, this was new territory to the majority. At that point, the main challenge was how to organize operations internally and externally with available resources. Later, when it became clear that the change was not short-term, more efforts went into the selection process of the best applications that would also be future-proof. Remote work became the primary form of work for most employees, and with an increased volume of work conducted with the use of collaboration platforms, employees started to discover advanced features of the platforms used. With remote work becoming more and more standard, employees' perception of collaboration platforms was affected. Research into user experience (UX) concepts can be conducted to better understand the acceptance of collaboration platforms. UX itself can provide an overall view of a certain product or service, taking into account the user's spectra of perceived feelings during familiarization and usage.

In this chapter, the digital collaboration will be explained with a focus on impacts on work and employees. Collaboration tools and collaboration platforms will be presented, and their characteristics and functionality features will be described. Collaboration platforms use will be analyzed, and collaboration tools and collaboration platforms will be shortly introduced. Further experience (UX) of Microsoft Teams, the collaboration platform will be analyzed. For the research study, we used a standard freely available UEQ questionnaire that is used for measuring the UX of interactive products and services. The scales within the selected questionnaire incorporate a wide-ranging sense of user experience. Standard usability aspects such as dependability, efficiency, and perspicuity as well as user experience aspects such as originality and stimulation, are assessed. The chapter will represent a comparative analysis of Microsoft Teams user experience by students of economics and business before COVID and in COVID time. The research aimed to identify the factors that influence students' UX when using the solutions mentioned during their study, where all classes have been shifted from blended learning (pre-COVID) to purely e-learning (COVID). The purpose of the research was also to identify changes in Microsoft Teams use according to changes imposed by COVID. In pre-COVID, Microsoft Teams was used for study material sharing and videoconferencing for lectures. Since Microsoft Teams offers other collaboration functionalities, such as sharing files, announcements, links, online quizzes, online contact hours via chat, assigning, and collecting assignments, many courses started to use these advanced functionalities in COVID time. Data collected based on questionnaires have been available from the first phase of the research conducted in times before COVID. We conducted the data collection using the identical questionnaire between COVID when Microsoft Teams was used in advanced level with additional functionality. The findings will be compared, and on the basis of comparative analysis, the changes will be discussed to summarize lessons learned.

4.2 DIGITAL COLLABORATION

In recent years, the phrase "digital collaboration" has been frequently used, especially in the talent management area and other business-related areas. The extent and different ways of collaboration are usually organization-specific, and it influences the portfolio of tools used to be efficient and successful (Eisenhauer, 2021). Collaboration is inevitable in modern business and is dependent on the sector, products, services, etc., of the organizations – some sectors require more innovative tools, some make do with more traditional toolsets. If we simplify, collaboration occurs every time two or more partners work towards a common goal. To be able to maximize the collaboration results of tools used, organizations must consider the reasons for digital collaboration (Eisenhauer, 2021). Many advantages come with digital collaboration, including the following:

1. **Saved Time:** Being able to finish a process or a task faster, automatically means saving time and, as such, means fewer costs for the organization.
2. **Strengthened Team Relationships:** One can look at teams in organizations as families. So, it comes to be very valuable to be capable of maintaining sound and effective relationships within groups of coworkers. With modern collaboration tools, this can be supported and elevated so that every member has a better understanding of teamwork and mutual goals.
3. **Better Organization of Work:** Collaboration tools are facilitators to improve the organization.

Collaboration can be on different levels, depending on the requirements of business processes. The challenge there is in knowing the process well enough, to be able to support it to the right extent with collaboration tools. Internal organization and external organization can be documented in great detail, and this can help in improving overall collaboration. To help this process, one should consider the next basic views:

1. **Team View:** At its core, teams are groups of 2 to 10 persons working on a common task. This type of collaboration is most commonly used, is, in general, the most effective, and brings the best results.
2. **Community View:** If the number of participating members is greater than in teams, we usually talk about communities. These can be organization-wide – so doing tasks that involve several internal departments or offices. So, one can be a team member and, at the same time, a member of an organization-wide community.
3. **Network View:** If collaboration includes members outside of the organization, we can describe the relationship as a network. We find this type of organization when external members are needed to do business processes (for example, consultants, customers, etc.). The main difference of network view, when compared to team and community, is in the fact that it connects members inside and outside of the core organization.

Digital collaboration can be described as a combination of tools and processes that enable partners, employees, and other participants to communicate and interact on different levels, and this is achieved with the use of select applications and systems (Deloitte, 2020). Digital collaboration tools for business are an effective way for organizations to be able to support different types of employees (working on location, working from home, working from anywhere) and still maintain effective business process flow. What we are seeing lately is innovation in digital collaboration, which is driving efficiency and productivity with the combination of traditional tools, social media, and other modern tools that create unique working environments (Madlberger & Raztocki, 2009).

Digital collaboration supports remote employees to be connected to others seamlessly, being able to complete their tasks and communicate. This positively impacts efficiency and enables better group dynamics and relationships. Digital collaboration is in some way progressive but is also getting more and more common. For innovative organizations, the challenge is in developing even newer, more advanced platforms and tools, to differentiate themselves and be leaders in their field.

The common workplace is changing, and the way we work is changing rapidly. This is due to the pandemic and has started even before that (Deloitte, 2020). With a growing mobile and distributed workforce, new generations of workers expect to collaborate anytime, anywhere. Innovative organizations are constantly searching and experimenting with new applications and tools that could help them in doing business. So the real challenge is with the developers, the vendors, to follow the trends and invest and place new tools on the market that workers can quickly adapt and use (McKinsey, 2020). For supporting global business, one must use tools that support virtual organizations and teams. In addition, new generations are coming to the workforce which are already familiar with remote collaboration technologies. This workforce is already experienced in using collaboration applications in their social life, so they expect to use similar or the same tools also in their work life – to be able to communicate and work regardless of location. One of their requirements (expectations) is flexibility. By some researchers, even by 2025, as much as three quarters of the active workforce will consist of millennials (Polycon, 2017).

Digital collaboration performance is affected by many factors (Eisenhauer, 2021). When analyzing efficiency in general or in terms of digital cooperation, one should take into consideration three elements:

- People;
- Processes; and
- Technology.

4.2.1 PEOPLE

People are in the foreground of every collaboration. People are collaborating, and with collaboration comes innovation and knowledge sharing. It is also important for organizations to lead by example – initiatives must come from leadership, and if this is done in the right way, one can expect increased engagement, improved morale, and productivity (Deloitte, 2013). The change in the way work is done also means that people need to adapt rapidly. As workers are compelled to work from home (WFH), their online behavior will change with time as they research new ways to carry out collaboration tasks. Because they don't have enough time to prepare, the quick change requires that new ways of doing this have to be found quickly, in addition to different restrictions (Waizenegger et al., 2020). There are a few issues associated with this particular element that organizations must consider:

1. **A Skilled Leader:** Digital collaboration does not function without proper personnel. Groups or teams work best when properly managed and led. This is typically done by a person or a small team that is skilled in management and leadership. But this usually is not enough. To be able to be effective, a leader would have to have previous experience with digital collaboration tools from previous projects.
2. **A Capable Team:** Having a competent manager is a good start, but we must watch out for the team in teamwork. All team members need to be experienced and knowledgeable in digital collaboration tools. They also must be flexible, with strong communication skills, team players, to contribute to the task at hand. Selecting team members is a very delicate task, even in a traditional setting. Due to the specifics of digital collaboration, managers must be very careful in this phase, considering the technological element.
3. **A Common Purpose:** Among the most critical characteristics of quality digital collaboration is the focus on the task at hand – all participants need to have the same goal. When a group of people

works together, with clear priorities, we can expect to achieve more in a shorter amount of time. Team members must feel connected and a part of that common goal – they will go that extra mile to contribute their fair share. It is important to document the goal right at the start.

4. **Reliance:** It is hard to reach, but at the same time is necessary for successful collaboration. Only when members of the team trust each other, will the teamwork be effective. Reliance is gained by experience, and managers should encourage a positive work environment with shared dedication and responsibility. Trust is gained faster in-person, so because this is not always possible in digital, different team building activities should also be organized, which are not strictly business-related.

4.2.2 PROCESSES

Processes, the second element of digital collaboration, are equally important and need to be considered. Basically, the process describes workflow used during collaboration. In most cases, processes need to be adjusted and changed from traditional to collaborative and need to be digitally supported. When changing one's process, some aspects of change should be considered so that we can ensure that you're getting the most out of digital collaboration:

1. **Strategy:** The basis is the strategy outline, which is commonly done by the team leader. Having a clear strategy helps team members in cooperation and is a tool that leads us to reach goals. This is especially true with collaborative tasks.
2. **Priorities:** After outlining the strategy, the team sets priorities so that everyone knows the order of things. In teamwork, every member usually has their own tasks, so also priorities are different for each team member. Some overlap is not uncommon. In general, priorities help us in following the outlined strategy and in keeping the focus on the goals.
3. **Data Distribution and Sharing:** For a team to be efficient, we must make sure that all team members have access to needed data and that data is distributed and kept always updated. When the

majority of the work is done online, usually some data collaboration space is available to facilitate data sharing. If teams are working with confidential or sensitive information, appropriate guidelines need to be set so that security is not affected.

4.2.3 TECHNOLOGY

Technology is another integral part of the collaboration and, basically, the one that facilitates remote collaboration and online work. When we think of the world today, compared to times over 10 or 20 years ago, technology has progressed in a staggering way. Digital collaboration tools can be overwhelming, but with all the social media we use every day, you probably won't have a problem with the use. Several technological platforms are capable of fostering digital collaboration, together with social intranet software (Polycon, 2017). Without social intranet software, digital collaboration would not be as simple as needed. Intranet technologies serve as a hub for employees to be able to share ideas, documents, organize meetings, and do other common tasks regardless of their own location. Digital collaboration tools are rapidly replacing technologies, like internet forums, bulletin boards, and online databases (Polycon, 2017). Social intranet software can be seen as a focal point of future digital collaboration, but other tools can also be used. Teleconference applications, VoIP telephone systems, and other tools all provide the possibility for people to work together, to have direct communication capabilities (voice, video, etc.), regardless of location. Although it depends on the sector and business process, a lot of organizations already use these platforms. For some, they are essential for everyday operations. Some applications combine collaboration features with teleconferencing features. Technologically also challenging is the need to still have access to more traditional applications and systems, like inventory, leads, etc., so organizations need to implement compatible systems that support the outlined strategy and the processes.

Collaboration has been an important aspect of successful organizations. And for the most part, in the past, employees were obligated to work at the employer's premises. When working from the field or on business trips, phone, and e-mail helped in maintaining communication and collaboration. The phenomenon of social network sites that at first was intended for manly personal use has evolved in time. We can define such networks as online services that let persons: (i) make an open or semi-open profile; (ii) create a list of users to share a connection with; as well as (iii) view and distribute their catalog of connections and the connections of others in the system. The landscape and organization of these connections usually vary between providers (Boyd & Ellison, 2007). As technology progressed, more advanced solutions came to the market — social intranet software with digital collaboration tools. This refers to the phenomenon of social networking in an enterprise context. As such, we can differentiate between two different types of enterprise social networks based on the involved actors. Firstly, we can refer to social networking on intranet social network platforms, which can functionally be compared to social networks but are only accessible in the enterprise intranet. Hence, the set of individuals is restricted to employees of an enterprise, and the links mostly reflect professional relationships. Secondly, we can refer to the phenomena of enterprise usage of public social networks. Examples are company pages on Facebook, where interaction takes place with potential customers, and business-related social networks (e.g., LinkedIn) for recruiting (Richter et al., 2011).

Social intranet software comes in different forms, but all tend to have strong collaboration features. With the use of such digital collaboration tools, organizations can help team members in working together in a truly global setting (Deloitte, 2013). If we compare different available social intranet platforms, there are differences and similarities, with some standard capabilities that include:

1. **Document Management System:** For business use, teams need to have the ability to share documents reliably and fast. Most platforms have strong document management support built in.
2. **Communication System:** Communication features of most platforms tend to replace traditional communication ways, like e-mail, etc. Since the features are all within the system, the use is simple and fast. One big advantage is also in the ability to access archived conversations.
3. **Content Creation Support:** Content is important, as it is a strong tool for attracting an audience both internally and externally. We can use third-party tools for creating content, but most platforms have modern content creation features that help users in creating stunning news posts, newsletters, marketing collateral, with SEO support, to engage our goal audiences easily and effectively.

4. **Advanced Search:** With increasing volumes of new data (files, blog posts, conversations, messages, etc.), users rely on the ability to quickly find legacy data using search. Platforms have search capabilities that help users in obtaining information in a user-friendly way.
5. **Advanced Calendar Features:** Time management is very important for employees. Calendar features are common in most traditional applications like MS Outlook or Lotus Notes. With the amount of time spent on social platforms, calendar features are integrated directly into applications. Some platforms have synchronization capabilities, so that meeting information is shared and synchronized across platforms and applications.

Organizations are facing the challenge of using collaboration tools to help drive business. The use of such tools for sales process support is common – this mostly comes in the form of CRM solutions. The larger the organization, the larger the impact on productivity and efficiency. But equally important is the use of these tools in small and medium-sized companies. It takes quite some time for an organization to fully implement and adjust to using social collaboration tools. In the case of large organizations, it can take up to 3 years. So, there is a gap between implementation and effective use of social platforms in the business. In the meantime, processes change. We put workflows in place and start measuring and monitoring impacts. And only at that point can we be efficient. We need to collect feedback from our audiences and analyze the data to change our upcoming activities. What we see in recent years is that some features introduced by social media, are being integrated into standard business information applications, like CRM, ERP, and others. This can be a source of innovation if organizations are capable of using the data in the right ways.

1. **Focus on Open Communication:** Communication is the basis of any collaboration. No matter the size of the team, members should be able to freely communicate, exchange ideas and concerns. Team leaders should promote an open communication environment because leading by example is usually the best practice.
2. **Organize Training:** In every implementation project, one of the important success factors is user training. Since social intranet software is typically easy to use, training is also quite

straightforward. If users don't have previous experience with the tool, the easiest way is to organize a training session that is aimed at covering basic features and scenarios. We can do this remotely or in a classroom. It is recommended that this kind of sessions are short, with the possibility to organize a set of short sessions that cover parts of features. This way, each member can choose those sessions that he finds useful.

3. **Identify Key Users:** Some employees tend to be more advanced and prone to picking up new applications faster than others. We can use this fact and include them in the process of implementation or appoint them as support persons. Also, other employees tend to look up to those users.

Intranet software offers a platform for social interaction among employees and the organization. Intranet offers collaboration tools, information sharing, messaging, document management systems, calendar support, etc. It is intended to support the positive work environment which provides to the global success of the organization.

Intranet software improves mobilization of assets, users, and stakeholders in being more creative in their duties. If we manage to integrate collaboration tools with basic business systems, like ERP or CRM, we can expect synergies that consequently mean increased productivity. The technology advances in recent years have enabled software vendors to integrate solutions in packages – this way, the data flow is easier, and the employees get a mutual interface to use.

Company culture is a critical part of the work environment and should be managed properly. If we encourage communication and collaboration over company-wide social systems, we achieve more efficient workflows, fewer obstacles, lean processes, and so on. Employees tend to appreciate a more relaxed atmosphere, they get empowered by the fact, that they can do more on their own using the platform at hand.

New generations of employees and users expect to have available similar collaboration tools that they already use in their free time. It goes the same for people in different roles, like consumers, business partners, employees, etc. With available technology, we don't have to limit business settings to strictly business tools. People expect changes and new ways of communication and collaboration quickly get adopted by innovative organizations. With the pandemic and the widespread of remote working, technology has made different scenarios possible. And to be honest – in the end, all that matters is that the tasks are done and the business is progressing. Traditional ways of work are not efficient anymore (apart from some specific sectors). Organizations are forced to adapt and use innovative approaches. We are nearing a time where home and work environments won't differ much. And if we turn the table, we expect no less also as customers. A customer expects personalized products and services, instant responses and communication with service providers and queries, etc. Depending on the size and volume of interactions, proper systems and tools must be used, to be able to support the ever-growing list of customers, employees, and partners.

COVID-19 pandemic is impacting the health sector severely and consequently also economies as a whole and organizations of all sectors. While the situation unfolds, organizations need to prepare for what is coming next. But not the next "normal." Normal is not available to us in most cases. The new reality is a combination of complexity, uncertainty, and opportunities. To adapt and thrive in these changed circumstances, companies need to become even more agile. They are forced to manage their everyday needs with longer-term opportunities while supporting the populations they tend to. Traditional places of work are changing. More and more organizations are reorganizing, downsizing office space, and redirecting their employees to work remotely from anywhere in the world. Virtual organizations are not really new, but the degree of virtual work has become overwhelming. These work settings need to be supported by the right connectivity tools. Workplace culture fulfills an individual's need for connection and belonging. It can improve morale by encouraging interpersonal support and guidance. These days, one of the deciding factors why people join organizations is because they have common values and share a larger purpose. A transparent, intentional culture is the basis for workers' sense of connection to something bigger than themselves. To shape a strong virtual culture, organizations need to take this vision and enable it remotely. Effective remote work is achieved through innovative technology tools, communication, and set policy. Organizational culture needs to promote confidence, create a sense of belonging, be managed from a human-centric point of view, and offer the tools for successful virtual work.

The COVID-19 pandemic has impacted the way employees of the majority of organizations work and cooperate. Performing tasks from home have come to be completely normal, and interactions have become virtual. Communication and collaboration have once again been impacted as more and more people are also returning to the workplace because of vaccination. But the changes that we have witnessed will not be undone – working hybrid or to some extent from home is the new standard. These changes will be permanent. All of this will result in a completely changed work organization, where some employees will be able to WFH, with others returning to employer locations.

Managers are being forced to further develop their changed businesses rather than returning to old patterns. Organizations are now working on implementing the innovations that they were forced to make, into their new workplace – they need to reinvent the models so that they stay productive in these post-pandemic times and also to be prepared for the next one. If anything, organizations have learned in the past couple of years that nothing is set in stone and that the world can change overnight. Some aspects of the reinvented workplace should be considered (Kane et al., 2021):

1. **Flexibility in Usage:** Organizational support and skill are important. Depending on the form of work being done, office space must be adjusted to better support the work process. Innovative organizations have invested in office space innovation even before the pandemic, now it is clear that different settings must be possible. If we are conducting an in-person workshop, doing a brainstorming session, or having a conference call – all these scenarios demand different office configurations.
2. **Virtual Work Evolution:** The switch from workplace to remote work has happened practically overnight. As the health situation is getting better, employees are gradually returning to the previous setup. A majority of organizations have had less than a year of experience with remote work and had not entirely recognized the good and bad aspects of it. Virtual working or some type of hybrid combination offers a lot of opportunities. With this switch, new collaboration tools have been put in place that offer new ways of communication and collaboration. Asynchronous work can bring benefits to organizations; they just need to understand them enough for this to happen. Also, the reality is that the most beneficial will be combinations of virtual work with more traditional work patterns. It is up to the management of organizations to reengineer their internal processes in such a way to benefit the

organization as a whole. Managers in organizations who still use hybrid meetings can innovate with the use of specific tools to include remote members more efficiently.

3. **Digital Support of Co-Located Work:** When employees work remotely from home, their focus is on their individual work tasks, or they participate in remotely organized meetings. If they have combined that with occasional or regular office workdays, this is mostly for tasks that require physical presence. What specific digital tools and applications support is that we can monitor who will be present in the office and when so that some meetings or tasks are organized accordingly. These kinds of applications support increased digitalization, and with their analytical features, organizations can improve internal workflows, workplace productivity, and time management.

New digital collaboration tools and applications facilitate new innovative work patterns. This has been and still is a continuing process, and some types of work will become automated in the future. Organizations should consider how tools of today could support the work process of the future years. Also, one important aspect for managers is the fact that more and more employees use hybrid work models with combining remote work with work from offices. The challenge for managers is in managing the workload and performances of employees.

Solution providers and collaboration tools vendors are constantly innovating and developing new products. The challenge for them is in the changes in the workplace. As we stated above, the share of the hybrid workforce has increased substantially, and vendors will have to adjust their solutions even more to support the changed scenarios. What can already be seen nowadays is the fact that the need for collaboration tools and teleconferencing tools is not diminishing with the return of the workforce to the offices. Even more – the usage of such tools is expected to even increase over the next years. Hybrid workplaces, hybrid teams are here to stay – global presence is easier than ever, and a lot of organizations are already benefitting from this new reality. With the help of different collaboration tools, organizations are no longer limited by borders and distance in their activities, being on the sales side or on the talent acquisition side. Vendors, who will keep innovating in this field, will have a vast market to cover.

Effectively combining virtual work and on-site work is quite complicated and comes with a lot of challenges. Some of them we have already mentioned above. Experience shows that hybrid work is especially associated with:

1. **Workforce Isolation:** The inclusion of remote workers is a big challenge. With working from home employees lack in-person contact and conversation, are less connected with day-to-day happening, don't fully connect with office culture and events. People are social beings and lack of social contact can negatively affect workers. Remote workers will have it harder with social contacts, organizational connectedness, and loyalty.
2. **Security Risks:** Security is easier managed within company walls. With remote work the possibility of cyber-attacks and security infringements grows as workers use their own devices, connect from their home networks or different locations. Modern collaboration tools offer a similar experience for those dialing in remotely and those working from the office. Security practices will have to be updated so that hybrid work models won't pose a greater threat than usual. Some collaboration tools have additional security features built-in so that in combination with standard security tools, the risks are manageable.

In conclusion, resilient organizations are creating technology-enabled connectivity for their contributors. There is a solution for every need, so your team can stay on purpose and thrive. While COVID has precipitated this reality, organizations can use the opportunity to apply proven digital strategies and create a sustainable virtual culture that can mark the difference between staying on course or falling behind. An innovative mindset in adopting a digital workplace experience will have lasting benefits, not only for your workforce, volunteers, and partners but also for the health and resilience of your organization and the recipients of your ever-important work.

4.3 COLLABORATION PLATFORMS

The offer of collaboration tools and solutions has intensified since the beginning of the pandemic. When we talk about collaboration, we mean this in the broad meaning of the word, considering internal collaboration among employees and external collaboration with our partners across the whole value chain. The reality is that several different applications need to be used to cover all aspects, as no universal application or software can cover all business scenarios. Non the less, some cover more, and some cover less. The development of collaboration tools and solutions is using past experiences to provide elevated functionalities for workforce engagement, increased productivity, with excellent security features Regardless of the vendor, collaboration tools must be in line with business needs. Applications and devices establishing corporate systems and cooperation can be found in different forms. Groupware is the commonly used expression for these kinds of systems. Others also use the term collaborative software or applications for computer-supported cooperative work. A mutual characteristic for all is that they deliver the collaboration modules of business processes.

The 3C collaboration model created by Borghoff & Schlichter (2000) is established on the belief that communication, coordination, and cooperation, are the most important issue for participants in a collaborative class. We can assign applications and tools to the different dimensions, but it is also clear that every tool supports other dimensions to some extent. The main characteristics by dimensions are (Broser et al., 2009):

1. **Communication:** This involves exchanging messages and the mediation of pledges and is mainly supported via video/phone conferences, e-mail, instant messaging, discussion lists, and the like.
2. **Cooperation:** It is the combined creation of group members within a common space, creating and handling cooperation items to achieve assignments. Supportive tools normally help via storage of documents, handling of applications, or message-keeping systems between companies.
3. **Coordination:** It facilitates people, resources, and activities, to resolve conflicts and simplify cooperation and communication. Coordination means providing for resources, scheduling, monitoring, human resources, and project management.

Each dimension is supported by a set of tools. But there is no tool to support all three dimensions adequately. Even though we can analyze particular dimensions individually, they ought not to be handled isolated, as there is a constant interaction between them. We should also consider that the dimensions and their tools concentrate on regulating and strengthening social contacts among team members, even cross-organizational.

The development of collaboration tools has been very rapid in the past few years. Collaboration has become essential as employees are no longer located all in one place but rather work in a hybrid way – some from remote locations, a part of the time on employer location. If organizations want to support this kind of work process, they need support in the form of collaboration tools and applications. These new and coming tools feature document management support over the cloud, video calls and meetings, time management support, and all of this on any device available (desktop, mobile, or tablet). The use of these tools positively influences productivity and improves efficiency. A lot of focus goes into making tools user-friendly, reliable, and easy to understand. If we manage to equip our workforce with the right collaboration tools, we will see empowered individuals capable of efficient problem solving and successfully following the business process workflows.

If we want to cover these requirements, the market is full of available tools. Grouping them is somewhat tricky, as applications usually have more than one purpose or functionality. Nonetheless, we can investigate unified communications as a service (UCaaS) solutions and meetings solutions in the basic sense of collaboration tools (Gartner, 2020a). UCaaS is identified as a cloud service that delivers several mutual functions as unified communications (UC) solutions on-premises (Gartner, 2020a). Core functionalities that UCaaS applications have are:

- Telephony includes enterprise telephony capabilities, voicemail, and unified messaging.
- Meetings include multiparty audio/videoconferencing with support for sharing content (files, applications, screen), along with messaging.
- Messaging capabilities offer support for text message interaction in real-time. They have evolved to cover two basic forms: one-on-one personal messaging and message exchange between group members and team members utilizing everyday user experience.
- Mobility and software clients. Solutions provide support for all forms of devices (desktop, web, mobile, tablets), and integrations for common business productivity tools, like calendars or e-mail applications.

Optionally capabilities include communications-enabled business processes. The ability to integrate communication features within other applications is appreciated. Different scenarios are potentially supported – from call centers, workgroup applications, plug-ins for cloud business information solutions, also integrations of artificial intelligence (AI) for business analytics.

Gartner recognized Microsoft Teams application as a leader for both UCaaS and solutions for meetings (Microsoft, 2021). Microsoft was placed highest of all solution providers for its ability to execute both reports. Microsoft Teams is an application that supports internal and external team members in connecting and collaborating synchronously (Microsoft, 2021). People can have meetings or calls one on one with fully integrated voice and video, informal chats, co-authoring a document, or work together in other apps and services. Microsoft Teams offers a shared workspace for people to iterate quickly on a project, work together with team files and collaborate on shared deliverables. Every new team created a new group, an online site using SharePoint complete with a document library, a shared mailbox and calendar (Exchange online), a OneNote notebook, and ties into other Microsoft 365 and Office 365 apps such as Planner and Power BI (Microsoft, 2021). Microsoft is rapidly developing new capabilities into Teams app – there is an update every month, and new features are added by the dozens. They also support the UserVoice initiative, with which everyone is able to suggest new features, and users vote on them – the most requested features get to be added in the coming updates. More recent added features incorporate (Microsoft, 2021):

- "Together mode," which offers a simulation of actual meeting members being in the same room;
- Customizable meetings that support setting up breakout rooms for meeting in smaller groups;

Ability to record the meetings on the go, accompanied with meeting notes and transcripts.

One of the latest investments by Microsoft is Microsoft Teams Rooms, which provide a mutual experience as in a standard office meeting room. This way, employees can experience hybrid work scenarios that support simplified meeting starts, effortless sharing of content with complete audio and video collaboration (Microsoft, 2021).

COVID pandemic has caused an increased demand for UCaaS, given that these tools are able to support unified communication services to users regardless of their location with little to no additional cost. On-premises business applications also support remote working environments, but this is often at the cost of additional licenses or even hardware requirements in the form of session border controllers (SBCs) at the boundary of company networks. UCaaS solutions facilitate increased user experiences for a diverse set of use cases (desktop, mobile), which raises the level of adoption. Increased demand and scale of remote working has forced organizations to search for and implement new collaboration tools with advanced features that also meet increased security and safety demands. These new features include, among others, background concealment, suppression of background noises, support for multiple users in simultaneous gallery scenes, and improvements to experience quality over unmanaged networks (Gartner, 2020a).

Meeting solution usage rose in 2020 due to COVID-related office closures and subsequent workplace policies that were put in place so that organizations could continue to perform business processes remotely. Solution providers were facing a challenge to provide services and tools for collaboration that could support businesses, educational organizations, and public sector agencies in the business. In many cases, the same applications were already in place but had to go into widespread use, as an overnight need for remote communication occurred. The pandemic crisis has hit the whole world, so leading vendors responded by offering some of their solutions for free for a limited period of time. This was also a good way for potential users to familiarize themselves with tools and features with the potential to become paying customers later (Gartner, 2020b).

What we could observe was a domino effect – collaboration tools became the main facilitator for any kind of communication – from business, education, public sector, healthcare, to even personal use as an additional way of staying connected with family and friends in times of limited movement or quarantine. A number of everyday services started to use these solutions only to stay in business – from religious services, yoga trainers, and other social events. All of them had to go online completely or partially, and collaboration platforms made all this possible. Gartner expects in coming years a positive effect in organization-wide use of videoconferencing as a part of the workforce will become regular users of remote meeting capabilities for work purposes regardless of location

(Gartner, 2020b). If we look at collaboration tools in a business setting, they support advanced interactions for teamwork, online presentations, training sessions, webinars, etc. Gartner also predicts new innovative use case scenarios in areas that didn't yet use these kinds of tools, for example, municipality council meetings, courtrooms, remote banking sessions, etc. What we can already observe is the fact that solution providers are increasingly expanding features and capabilities that support specific use cases. This can be done by goal-oriented development or by integrating other vendor solutions via plug-ins.

Predictions are that different collaboration platforms and tools will be used even more than today due to the COVID pandemic and its consequences to the global economy. Some predictions show that 74% of organizations will change at least 5% of their usual on-site staff into permanent working from home positions by 2022 (Gartner, 2020b). Another prediction is that due to the changes in workforce demographics and due to the amount of working from home, by 2024, only 25% of enterprise meetings will continue to be conducted in person, as opposed to this share being 60% today (Gartner, 2020b).

Collaboration platforms development intensifies with increased user volumes, resulting in expanded feature lists. Furthermore, as new features are rolled out, the users get more acquainted and familiar with the platform and start using advanced features. This process, in time, influences users' attitudes toward collaboration platforms and their perception of the new technology used. User experience (UX) concepts can be used in research on platform acceptance. There are a few common features that we find (and need) in an online collaboration tool:

1. **Instant Messaging:** The ability to message anyone is a must-have feature. Also being able to have group messaging with different prompts. In this regard, maybe e-mail seems outdated but remains the main communication channel in business. Collaboration tools usually support group and one-on-one messaging, as well as a specific channel or tag-based messaging.
2. **Video Conferencing:** Whether you're scheduling a virtual meeting, or need to chat with a team member on the fly quickly, video conferencing is a must-have.
3. **Document Sharing:** Some platforms allow document management and sharing, sharing of cloud-based files as links, from cloud storage like OneDrive, Google Docs.

4. **Project Planning and Tracking Tools:** We can use platforms that focus on communication; others focus on project planning and tracking. Depending on demands and the fact that a lot of work is done over projects, this is quite a useful feature.
5. **Reporting and Dashboards:** Progress reports and monitoring is a must – built-in features enable out-of-the-box reporting or at least export features to analytical tools that support advanced analytics and dashboarding.
6. **Sharing Permissions:** The ability to flexibly manage sharing of files and folders is greatly needed. Platforms offer a variety of different sharing rules and access features so that we can support any business scenario possible.

So, the question is also, to what extent the functionality is being used, and, with time, does the use of collaboration platforms, such as Microsoft Teams, change the work patterns of users. Some researchers have already started noticing this with their studies (Lansmann et al., 2019). Hence, as with our research, it is interesting to compare the data from different time points using the application and to compare results.

4.4 USER EXPERIENCE

When we evaluate software applications, different aspects can be used. Some are quantitative, but it is often down to the user's subjective opinion, whether he or she finds an application good or not. One characteristic often used is User experience (UX), which relates to the end users' actual experience with the software. ISO/IEC 9241 defines UX as "a consequence of the presentation, system performance, functionality, interactive behavior, and assistive competencies of an interactive system, both software and hardware. It is also a consequence of the user's prior experiences, skills, attitudes, personality, and habits (ISO, 2010)." It is restricted to products and entire systems and services with all that fits the user voyage creating a user experience before utilizing a product or system (Van de Sand et al., 2020). Good UX contributes to higher work motivation and performance and can also affect the well-being of users, and is crucial to maintain or gain market shares (Hassenzahl, 2010; Nass et al., 2012). It is reasonable to anticipate that a good UX will lead to greater satisfaction levels, which can guarantee application success. UX is most frequently connected with products and programs, concentrating on making sure end users have a clear and useful experience with solution interface. User experience is greater than the user interface (UI). A product can not only be viewed in a single way; users do not have experiences only at the time of use. Users' perception is also influenced by the accompanying service and the entire system of a product (Van de Sand et al., 2020). Usability is often seen as a necessary precondition for good UX (Lallemand et al., 2015). One of the widely used definitions of usability is given by ISO/IEC 9241: "the degree to which a system, service or product can be used by particular users to achieve particular goals with efficiency, effectiveness, and satisfaction in a particular context of use (ISO, 2010)." UX has five unique characteristics that differentiate it from usability (Hassenzahl, 2010):

- Subjective (heavily relies on human perception);
- Holistic (includes both hedonic and pragmatic aspects of use);
- Dynamic (changes over time);
- Context-dependent (is situated in context); and
- Worthwhile (encompasses positive and meaningful consequences of use).

The terms UX and usability are interrelated and often used interchangeably to measure whether the user requirements are satisfied. There is some significant distinction between the two terms. UX is concerned with the complete experience of a user while they use a system that is more prone to their emotional views. Usability is evaluating the excellence of their usage of a system based on efficiency, effectiveness, and satisfaction measures. To sum up, UX can be evaluated in relation to reaching different hedonic aims such as emotion, human behavior, and response that is generated based on interactive experience. Usability is also evaluated in relation to reaching some kind of performance objectives such as accessibility, safety, learnability, and similar (Zaki & Nazrul Islam, 2021).

For a positive user experience, it is the designers' objective that the intended nature of the product generates interest, enjoyment, and gratification. From a user point of view, abilities are apparent, estimated, and experienced in the perspective of usage, which ideally leads to interest, enjoyment, and gratification. Still, this can only be accomplished by a particular degree of pragmatic and hedonic qualities. The substance and usefulness of a product need to be sufficient and beneficial. Interactions need to be easy to comprehend and effortless. The presentation must be attractive, pleasing, and in unity with the brand character (Van de Sand et al., 2020).

Standards or agreed-upon lists of principles and practices that can support delivering or identifying good UX are still not available. Scattered examples of UX practices and principles can be found in UX literature. Table 4.1 tabulates some descriptions and examples (Kashfi et al., 2019).

TABLE 4.1 The Definitions of UX Principles, Practices, Tools, and Methods

Term	Description	Examples
UX principles	Important factors and fundamental concepts that reflect the understanding of UX as a phenomenon. Practitioners need to take these principles into account in their work	Both hedonic and pragmatic aspects of software use play an important role in forming UX; UX is temporal.
UX practices	Activities that practitioners need to perform to satisfy the principles.	Identify users' personal goals and preferences, create prototypes, involve users in the design process, evaluate the software from pragmatic and hedonic perspectives.
UX methods	Impose structure on the practices to make them systematic and ultimately more likely to be successful.	The survey, questionnaire, mind mapping, field study, cognitive mapping, design studio.
UX tools	Computer-based programs or analog means that assist practitioners in performing various practices and are often designed to support particular methods and intend to make practitioners' work more systematic.	Persona, eye-tracking programs, visual design, and prototyping tools.

Source: Adapted from Kashfi et al, 2019. © 2019 Elsevier.

UX evaluation is a popular topic in studies, and many assessment techniques have been created in the past. As we have shown, how difficult it is to define UX is even more difficult to evaluate and measure. Before their first encounter, people can have indirect experience through anticipations based on existing understanding of associated technologies, brands, advertisements, presentations, demonstrations, or others' opinions. Similarly, the indirect experience extends after usage, such as reflection on the previous usage, or through changes in people's appraisals of use (Roto et al., 2011). One important aspect to consider is also timeframe. We can argue that UX changes over time. At one extreme, we can focus solely on what someone has experienced for a very brief moment during usage.

On the other, we can focus on aggregated experience developed through a string of usage encounters and phases of non-use that might span over longer periods of time. UX can consequently indicate a specific impression during use (momentary UX), a judgment of a specific usage episode (episodic UX), or opinions about a complete system after using it for a while (cumulative UX). Anticipated UX could correlate to the time before first use. A summary of different types of UX is shown in Figure 4.1 (Roto et al., 2011).

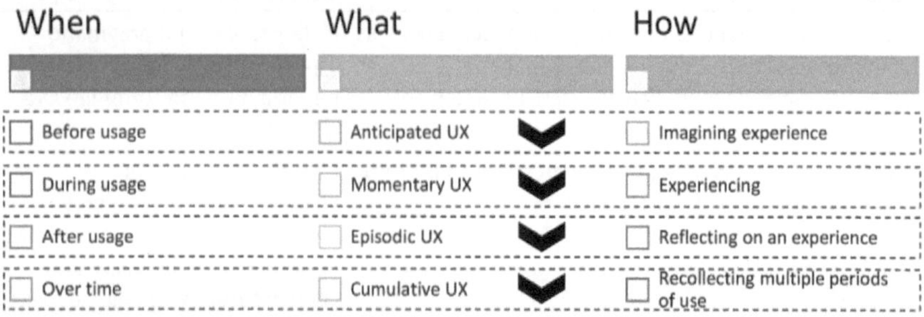

FIGURE 4.1 UX in reference to time.

Source: Adapted from Roto et al., 2011

There is a variety of methods specifically created to assess and research UX-related concepts. For the research study, we used a standard freely available UEQ questionnaire used for measuring UX of interactive products and services, which is described in succeeding sections.

4.5 METHODOLOGY AND RESEARCH

Synchronous, different places communication and collaboration must work smoothly, now more than ever due to the ongoing pandemic in the world (Stoytcheva, 2018). Since Microsoft Teams is a rather complex system, it demands its usability (Bergman, 2021). To examine the usability of a system, one needs to study how efficient, effective, and satisfactory a product is. In addition to this, it is important to address a specified context of use by a specified user with a specified goal. As mentioned before, we decided to use a standard freely available UEQ questionnaire. The authors of the UEQ developed a questionnaire that allows a quick assessment, including a rather thorough sense of user experience. It should permit the users in a very simple and immediate way to convey emotions, thoughts, and opinions that occur when interacting with the product in question (Laugwitz et al., 2008). The UEQ contains six scales containing 26 items (Laugwitz et al., 2008):

- Attractiveness (General sense of the product. Do consumers like or dislike the product?);
- Perspicuity (Is it easy to get acquainted with the product? Is it easy to learn to use the product?);
- Efficiency (Can consumers work out their tasks without needless struggle?);
- Dependability (Does the consumer think he controls the narrative?);
- Stimulation (Is it stimulating and inspiring to use the product?);
- Novelty (Is the product modern and innovative? Does the product catch the attention of users?).

Attractiveness is a genuine valence component, efficiency, perspicuity, and dependability are pragmatic quality aspects (goal-directed), whilst stimulation and novelty are hedonic quality aspects (not goal-directed) (Laugwitz et al., 2008). Every dimension is consisting of more scales which is shown in Figure 4.2.

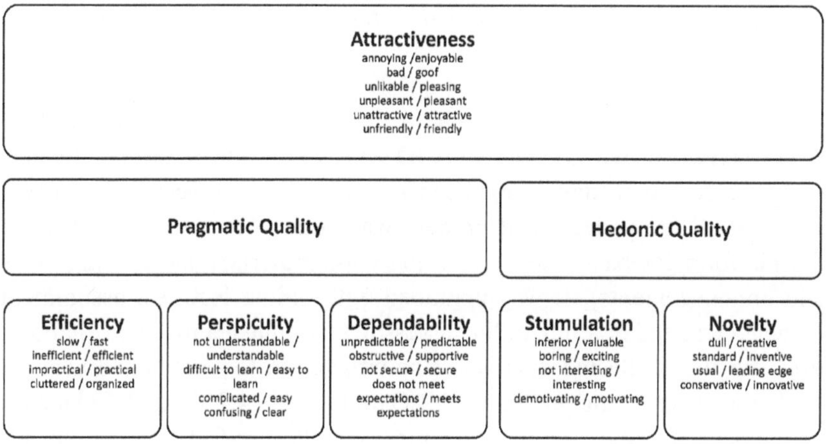

FIGURE 4.2 UEQ scale structure.

We have conducted a comparative analysis of Microsoft Teams user experience by students of economics and business based on pre-Covid and

Covid sample data. The research focus was on determining factors influencing students' UX of using collaboration solutions available during their study. A fundamental change in the study process was caused by the Covid pandemic in a way, that all the courses were being delivered in the form of pure e-learning in the Covid period, while blended learning was used predominately in the pre-Covid period. Moreover, in the pre-Covid period, Microsoft Teams was used to distribute study material and deliver video lectures. In the Covid period, when the shift to pure e-learning occurred, other advanced functionalities of Microsoft Teams started to be used, for example, sharing files, announcements, links, online quizzes, online contact hours via chat, distribution and collection of assignments etc.

Data collected based on questionnaires have been available from the first phase of the research we conducted in times before COVID. We repeated the data collection using the same questionnaire between COVID when Microsoft Teams was used in advanced level with additional functionality. Pre-COVID research was conducted with paper questionnaires during the winter semester among the students who had no previous experience with Microsoft Teams. Microsoft Teams was used as a platform for sharing class materials and for sharing news and announcements. We were able to get 147 responses (sample size), which were then transferred to Microsoft Excel for further analysis using predefined UEQ tools. COVID sample was smaller due to different reasons – mainly because of less motivated students who were already in almost a whole year of distant learning. Also, this time we used electronic questionnaires, which are easier to use, but as there is no control over the actual process of filling in the data, the response was weaker. Our COVID sample was 44 responses. Students gave answers with daily, advanced use of Microsoft Teams due to the fact that all the pedagogical process was done online. On top of using the platform for basic means (sharing class materials and sharing news and announcements), it was upgraded with online lectures, assignment management and grading, online assessment, online teamwork, collaboration, etc. The collected data was also transferred to Microsoft Excel for further analysis, using predefined UEQ tools. Table 4.2 shows UEQ means of the scale's values for both samples. For the pre-COVID sample, we can see evaluation for all the scales, with the highest values on scales of Novelty, Efficiency, and Dependability. The lowest means value was with the Stimulation scale. When we compare the means result with the COVID sample, we can see increased values across all the scales. We can observe some changes in the highest and lowest values being that now the highest values are with Perspicuity, Novelty, and Attractiveness scales. The lowest means value is still with the Stimulation scale.

TABLE 4.2 UEQ Means of the Scales

	Pre-COVID	COVID
Attractiveness	0.913	1.280
Perspicuity	0.999	1.403
Efficiency	1.240	1.131
Dependability	1.049	1.182
Stimulation	0.391	0.489
Novelty	1.417	1.295

Figure 4.3 shows the UEQ means of the scales in comparison with confidence level intervals. The confidence interval is a measure of the precision of the estimation of the mean. The smaller the confidence interval, the higher is the precision of the estimation, and the more we can trust the results. We can observe that confidence intervals are bigger with the COVID sample than with the pre-COVID sample. This can mainly be explained by the difference in sample size, which was smaller than the COVID sample.

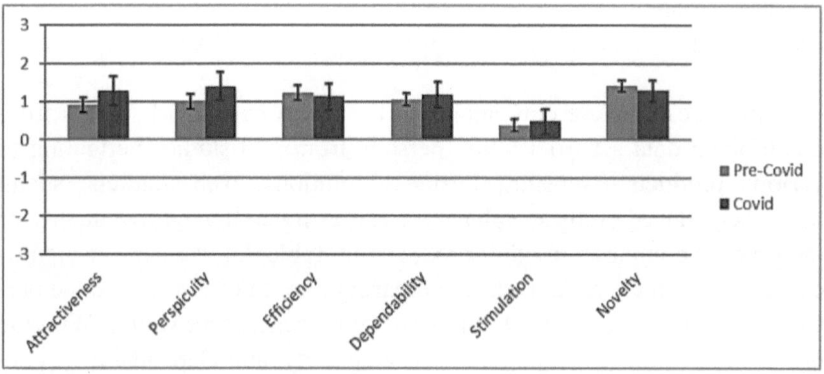

FIGURE 4.3 UEQ means of the scales.

Attractiveness is a pure valence dimension. UEQ scales can be clustered into pragmatic quality (Efficiency, Perspicuity, Dependability) and hedonic quality (Novelty, Stimulation). Pragmatic quality describes task-related quality aspects, hedonic quality the non-task related quality aspects. Efficiency, Perspicuity, and Dependability are pragmatic quality characteristics (goal-directed), while Stimulation and Novelty are hedonic quality characteristics (not goal-directed). If we look at the results in Table 4.3, with the pre-COVID sample, all mean values are positive, and Pragmatic Quality values are higher than Hedonic Quality values. When comparing results with COVID data, we can see that Attractiveness has gained a significant bump up compared to the pre-COVID sample. Also, Pragmatic Quality has increased; on the contrary, Hedonic Quality has slightly decreased. Due to the fact that there is a difference in the way that Microsoft Teams was used in the pre-COVID and COVID samples (basic vs. advanced), we can correlate this to the changes in Pragmatic and Hedonic Quality. Since Pragmatic quality describes task-related quality aspects, due to advanced and more frequent use of the platform, these values are higher in the COVID sample. On the other hand, Hedonic quality is practically the same or slightly decreased since there was no notable stimulation or novelty effects by increased use of Microsoft Teams.

TABLE 4.3 Pragmatic and Hedonic Quality

	Pre-COVID	COVID
Attractiveness	0.91	1.28
Pragmatic quality	1.10	1.24
Hedonic quality	0.90	0.89

UEQ questionnaire data analysis also offers benchmark values from a combined data set from 9,905 persons from 246 studies pertaining to various products (websites, business solutions, web markets, social networks). Based on these values, we can evaluate the relative quality of the product compared to other products. In Table 4.4, we can see significant changes in the results when comparing our scale values from the pre-COVID sample and COVID sample with the benchmark values. With the pre-COVID sample, Attractiveness, Perspicuity, and Dependability were below average (50% of results were better, 25% of results were worse). Stimulation was even marked as bad (in the scope of the 25% weakest results). Efficiency got above average rating (25% of results were better, 50% of results were worse), and Novelty scale was marked as exceptional (In the scope of the 10% greatest results). COVID sample benchmark comparison in Table 4.4, shows above average values with Attractiveness, Perspicuity, Efficiency, and Dependability scales (25% of results were better, 50% of results were worse). The Stimulation scale value is still bad (In the range of the 25% worst results), although slightly better. Novelty values are Good (10% of results were better, 75% of results were worse).

TABLE 4.4 Benchmark Comparison

	Pre-COVID		COVID	
	Mean	**Compared to Benchmark**	**Mean**	**Compared to Benchmark**
Attractiveness	0.91	Below average	1.28	Above average
Perspicuity	1.00	Below average	1.40	Above average
Efficiency	1.24	Above average	1.13	Above average
Dependability	1.05	Below average	1.18	Above average
Stimulation	0.39	Bad	0.49	Bad
Novelty	1.42	Exceptional	1.30	Good

If we summarize the comparison, we see that with increased use and advanced use of Microsoft Teams, the values with COVID the majority of scales show above average reading with the exception of stimulation. So, we could conclude that based on the benchmark values, Microsoft Teams platform UX is above average by the majority of the scales measured.

4.6 CONCLUSION

The COVID pandemic has brought tremendous challenges in all areas of our lives. The health aspects are the most important, but as time passes, we see changes in other areas. One of the more impacted areas is the way we teach, work, and do business. One common component in these areas is collaboration. We cooperate with each other constantly, and technology has facilitated new ways of collaboration that in the past were not possible. Groupware applications began to emerge with the widespread of the internet. These applications, which first supported basic communication, soon developed into whole collaboration platforms that include a wide variety of basic and add-on functionality. Cloud computing has also been an important facilitator for the success of some applications. Since there are low entry costs, everyone can try and use these applications. One of the top cloud providers is Microsoft, whose cloud efforts have made a massive accomplishment in the previous years. The COVID pandemic has brought challenges, but the infrastructure was in place to more or less cover the vast influx of users practically overnight. Microsoft Teams has proven itself as a reliable and capable collaboration platform.

As stated before, collaboration platforms have been around for quite some time. In higher education, we observed that Microsoft Teams used a supporting platform with basic functionality, like study material sharing and video conferencing. Later, because of online-only study, more functionality started to be utilized. Because of that, users spent more time using Microsoft Teams, and the user experience changed.

We have started researching user experience (UX) before, as it influences the way users perceive applications and products. In our daily lives and routines, we can observe that applications and products that have the same or similar functionality tend to be perceived differently by users. So, if one tends to develop a product or a service, UX is one aspect to consider. UX research is quite popular, and there is plenty of techniques in use. It's a fact that some facets of UX are subjective, and some are objective, so we have to have appropriate measures in place to make sense of the data collected. In our research, we have chosen the UEQ questionnaire, which measures the sense of user experience. Both traditional usability characteristics (efficiency, perspicuity, and dependability) and user experience characteristics (novelty, stimulation) are measured. The collected data shows that the COVID pandemic has impacted the UX of Microsoft Teams with our students. The Covid period has resulted in a more extensive use of various functionalities. This has resulted in increased UX values of measured scales. Also, a positive change was observed in the general UX perception. These changes are best seen in the case of the Attractiveness scale, where we observed the most considerable increase as opposed to pre-COVID data. We have seen similar results with pragmatic quality results, whereas hedonic quality, numbers have decreased.

We see a lot of possibilities for future research. On the one hand, we should gather more data with bigger sample sizes, and we could do a more reliable analysis. Also, one of the aspects of UX research in the time component – UX changes with time. The more hands-on time the users have with a product or a service, the more accurately they can access their experience. So, additional data could be gathered with users that now have used Microsoft Teams for the second year. Also – as some types of work will probably be permanently shifted to remote collaboration platforms, this is also one of the aspects we can consider for the future.

KEYWORDS

- **artificial intelligence**
- **collaboration platforms**
- **digital collaboration**
- **Microsoft Teams**
- **session border controllers**
- **unified communications**
- **user experience**

REFERENCES

Bergman, E., (2021). *Microsoft Teams: A Qualitative Usability Study (Dissertation).* Retrieved from: http://urn.kb.se/resolve?urn=urn:nbn:se:kau:diva-82356 (accessed on 7 February 2023).

Borghoff, U. M., & Schlichter, J. H., (2000). *Computer Supported Cooperative Work: Introduction to Distributed Applications.* Springer-Verlag New York, Inc., Secaucus, NJ, USA.

Boyd, D. M., & Ellison, N. B., (2007). Social network sites: Definition, history, and scholarship. *Journal of Computer-Mediated Communication, 13*(1). Retrieved from: https://onlinelibrary-wiley-com.ezproxy.lib.ukm.si/doi/epdf/10.1111/j.1083-6101.2007.00393.x (accessed on 7 February 2023).

Broser, C., Fritsch, C., Gmelch, O., & Pernul, G., (2009). Impact analysis of collaborative platforms – the case of SPIKE. *IFAC Proceedings Volumes* (Vol. 42, No. 25).

Deloitte, (2013). *Digital Collaboration: Delivering Innovation, Productivity and Happiness.* Retrieved from: https://www2.deloitte.com/content/dam/Deloitte/se/Documents/technology-media-telecommunications/deloitte-digital-collaboration.pdf (accessed on 7 February 2023).

Deloitte, (2020). *Remote Collaboration Facing the Challenges of COVID 19.* Retrieved from: https://www2.deloitte.com/content/dam/Deloitte/de/Documents/human-capital/Remote-Collaboration-COVID-19.pdf (accessed on 7 February 2023).

Eisenhauer, T., (2021). *Grow Your Business with Collaboration Tools.* Axero Solutions. Retrieved from: https://info.axerosolutions.com/grow-your-business-with-collaboration-tools (accessed on 7 February 2023).

Gartner, (2020a). *Magic Quadrant for Unified Communications as a Service, Worldwide*. Retrieved from: https://app.hushly.com/asset/content/cvd9stumsb74q0or34sh1trhkh (accessed on 7 February 2023).

Gartner, (2020b). *Magic Quadrant for Meeting Solutions*. Retrieved from: https://www.nuvias-uc.com/wp-content/uploads/2020/10/Full-Gartner-Report-2020.pdf (accessed on 7 February 2023).

Hassenzahl, M., (2010). Experience design: Technology for all the right reasons. *Synthesis Lectures on Human-Centered Informatics, 3*(1).

ISO, (2010). *ISO 9241-210:2010 Ergonomics of Human-System Interaction Part 210: Human-Centered Design for Interactive Systems*. Retrieved from: https://www.iso.org/standard/52075.html (accessed on 7 February 2023).

Kane, G. C., Nanda, R., Phillips, A., & Copulsky, J., (2021). Redesigning the post-pandemic workplace. *MIT Sloan Management Review (Spring 2021)*. Retrieved from: https://sloanreview.mit.edu/article/redesigning-the-post-pandemic-workplace/ (accessed on 7 February 2023).

Kashfi, P., Feldt, R., & Nilsson, A., (2019). Integrating UX principles and practices into software development organizations: A case study of influencing events. *Journal of Systems and Software, 154*. Retrieved from: https://www.sciencedirect.com/science/article/pii/S0164121219300731 (accessed on 7 February 2023).

Lallemand, C., Gronier, G., & Koenig, V., (2015). User experience: A concept without consensus? Exploring practitioners' perspectives through an international survey. *Computers in Human Behavior, 43*. Retrieved from: https://www.sciencedirect.com/science/article/pii/S0747563214005718 (accessed on 7 February 2023).

Lansmann, S., Schallenmüller, S., & Rigby, M., (2019). *Teams Everywhere – Investigating the Impact of Microsoft Teams on Knowledge Worker*. Retrieved from: https://www.researchgate.net/profile/Simon-Lansmann-2/publication/338038787_Teams_Everywhere_-_Investigating_the_Impact_of_Microsoft_Teams_on_Knowledge_Worker/links/5dfb67b7a6fdcc28372c8738/Teams-Everywhere-Investigating-the-Impact-of-Microsoft-Teams-on-Knowledge-Worker.pdf (accessed on 7 February 2023).

Laugwitz, B., Held, T., & Schrepp, M., (2008). *Construction and Evaluation of a User Experience Questionnaire*. Retrieved from: https://www.researchgate.net/publication/221217803_Construction_and_Evaluation_of_a_User_Experience_Questionnaire (accessed on 7 February 2023).

Madlberger, M., & Raztocki, N., (2009). Digital cross-organizational collaboration: Towards a preliminary framework. *Fifteenth Americas Conference on Information Systems*. Retrieved from: https://ssrn.com/abstract=1477527 (accessed on 7 February 2023).

McKinsey, (2020). *The Next Normal: The Recovery Will be Digital*. Retrieved from: https://www.mckinsey.com/~/media/mckinsey/business%20functions/mckinsey%20digital/our%20insights/how%20six%20companies%20are%20using%20technology%20and%20data%20to%20transform%20themselves/the-next-normal-the-recovery-will-be-digital.pdf (accessed on 7 February 2023).

Microsoft, (2021). *Welcome to Microsoft Teams*. Retrieved from: https://docs.microsoft.com/en-us/microsoftteams/teams-overview (accessed on 7 February 2023).

Nass, C., Adam, S., Doerr, J., & Trapp, M., (2012). Balancing user and business goals in software development to generate positive user experience. In: *Human-Computer Interaction: The Agency Perspective*. Springer-Verlag Berlin Heidelberg.

Polycon, (2017). *Are Your Collaboration Tools Helping or Hurting Your Organization?* Retrieved from: http://www.polycom.com/content/dam/polycom/common/documents/guides/collaboration-tools-benchmark-report-guide-enus.pdf (accessed on 7 February 2023).

Richter, D., Riemer, K., & Vom, B. J., (2011). Internet social networking. *Bus. Inf. Syst. Eng., 3.* Retrieved form: https://doi.org/10.1007/s12599-011-0151-y.

Roto, V., Law, E., Vermeeren, A., & Hoonhout, J., (2011). *User Experience White Paper. Outcome of the Dagstuhl Seminar on Demarcating User Experience, Germany.* Retrieved from: https://experienceresearchsociety.org/wp-content/uploads/2023/01/UX-WhitePaper.pdf (accessed on 7 February 2023).

Stoytcheva, M., (2018). Students' perceptions of online collaboration in a distance learning French language course. *AIP Conference Proceedings 2048*, 020030. Retrieved from: https://doi.org/10.1063/1.5082048.

Van De, S. F., Frison, A. K., Zotz, P., Riener, A., & Holl, K., (2020). The intersection of user experience (UX), customer experience (CX), and brand experience (BX). In: *User Experience is Brand Experience*. Management for Professionals. Springer Nature, Cham., Switzerland.

Waizenegger, L., McKenna, B., Cai, W., & Bendz, T., (2020). An affordance perspective of team collaboration and enforced working from home during COVID-19. *European Journal of Information Systems, 29*, 4. Retrieved from: https://doi.org/10.1080/0960085X.2020.1800417.

Zaki, T., & Nazrul, I. M., (2021). Neurological and physiological measures to evaluate the usability and user-experience (UX) of information systems: A systematic literature review. *Computer Science Review, 40*. Retrieved from: https://www.sciencedirect.com/science/article/pii/S1574013721000150 (accessed on 7 February 2023).

CHAPTER 5

The Culture Shock of COVID-19: A Study to Understand the Effects of Pandemic on the Art and Culture Industry

CHIRMI ACHARYA

Assistant Professor, Department of Psychology, Manipal University Jaipur, Rajasthan, India

ABSTRACT

When Winston Churchill was asked to cut arts funding in favor of the war effort, he simply replied, '*then what are we fighting for*?'

COVID-19 pandemic has been nothing less than the giant of all wars humans have fought lately, and it has had an adverse effect on the art industry across the world as well as in India. Cultural and creative industries especially performing arts have been the worst hit by the crisis. Owing to its nature as a cultural practice that requires the physical presence of an audience – no longer permissible under the pandemic – performance activities were among the first to cease. The Performing Arts industry plays an important role from many perspectives, including its contributions to the economy and society, employment, sustainable development, tourism, education, and a range of well-being benefits. But during the pandemic, the industry was worst hit. It has been very difficult for the industry and artists to survive amid the lockdown and many families whose soul work is their engagement with performing arts had a difficult time surviving with basic needs. Many artists have been affected economically and psychologically. The pandemic brought drastic changes to the art and culture industry

Reshaping the Business World Post-COVID-19: Management Strategies for Sustainable Behavior Change. Arvind K. Birdie & Ruchi Joshi, (Eds.)

with new normal being adopted in creative sector too, where concerts and performances were staged online to many artists bearing the brunt of having no means to survive. This chapter is an attempt to understand the social-psychological effects of pandemic on art and culture industry and artists in India and across the world and will also try to understand the coping strategies being used by the artists and how governments and culture organizations can nurture the art and culture industry survive from being drowned by the cyclone called COVID-19 pandemic.

5.1 INTRODUCTION

The year 2020 was historical as it put the entire world and human race on a standstill. With the COVID-19 pandemic raging across borders it brought human life to a slowdown. Every aspect of human life was affected by the pandemic, including economy and psycho-social aspects of human life. People across the countries were encaged in their homes and adopted the new normal at every front, be it education, business, and work space, as well as building a resilient act to combat with the catastrophic situation by making necessary changes and adjustments.

Creative, cultural, and performing arts industries have also been hit by the pandemic, the sector which is the heartbeat of any culture and civilization. UNESCO (2020) predicted that although the cultural sector plays an important role in building resiliency in times of crisis, but it will be the most hard-hit sector with respect to economic indicators. Art and cultural sector are unimaginable without an in person contact and this pandemic came as a shock to it. In the past decades creative and culture industry has become an important sector of the global economy too, and owing to the pandemic it has suffered huge loss (Li, 2020; Rodríguez-Gulías et al., 2020).

COVID-19 crisis affected the various fronts of creative industry which includes films, advertising, radio, museum, fashion sector and professions like musicians, dancers, performers, designers, and even the tourist professionals (Khylstova, Kalyuzhnova, & Belitski, 2021). Theaters, rehearsal spaces, studios, cultural centers were closed which impacted the artists, performers, and also the allied sectors which benefit directly or indirectly from this industry such as electricity men and tourism and it also had a detrimental effect on mental well-being of the artists and the ones who benefit from the positive effects of the industry.

5.2 EFFECTS OF COVID-19 PANDEMIC ON CREATIVE, ART, AND CULTURE INDUSTRY AT A GLOBAL LEVEL

"*Sorry, we are closed*" was written and posted all over the cultural centers across the world, at museums, libraries, auditoriums, art centers and studios. The immediate priority in times of crisis was to shut down the art activities as per the guidance of WHO, and the paradox was that there was no rule book to guide how to continue to pay artists and employees of cultural centers. Art is an emotionally moving experience for many and it requires one to be in a space with many people to experience the pleasure, but the irony of the time is such that art institutions were challenged with the need of the time to close their space to curb the spread of the deadly virus declared as a pandemic by WHO (Liscia, 2020).

All the museums in the USA like the Smithsonian, Brooklyn Museum, Harvard Arts Museum, Yale University Art gallery and other such places of importance closed down with immediate effect and also devised contingency plans to keep providing salaries to their employees during the time they were to remain closed (Kenney, 2020).

Organizers of various music festivals, concert coordinators and choreographers had to rethink whether to hold their shows or not as the tide didn't seem to stop affecting the artists and the allied professionals both economically as well as emotionally. The cancellations and rescheduling lead to confusion and chaos among the audience as well as the organizers who were already taking the brunt of huge economic loss, along with that a grave existential crisis was now looming over their heads (Pitchfork, 2020).

In light of social distancing to curb the spread of coronavirus, theater, and dance performances were also postponed as large gathering were only allowed at hospitals of medical centers amidst the fear of ever spreading deadly virus.

Creative industries all over the world have suffered many short term and long-term challenges like cancellation of events, bankruptcy, not enough money to pay the artists and employees, all of which has led to major social-economic-psychological repercussions (OECD, 2020). Many art museums, galleries, media houses and many art venues closed permanently due to the restrictions put on the sector by the governments owing to the pandemic (Beseley, 2020; Watson & Herman, 2020).

The first quarter of 2020 saw only closures and cancellations as the creative, art, and cultural industry rapidly restricted its activity due to sudden restrictions put due to the pandemic and this led to utter chaos as the dates for reopening and restarting the museums, galleries, film/TV shooting, concerts, etc., remained uncertain. Public expected the creative and art industry to become a source of resilience in the tough times, but in-person activities could not be started but the looming fear of the virus didn't allow it to happen early on in the pandemic.

Entertainment has been a source of hedonic pleasure and psychological well-being since time immemorial (Rieger, Reinecke, Frischlich & Bente, 2021). Be it traditional forms of music, art, dance, drama, or the modern mediums of entertainment in the form of radio, films, television, and in the recent years' digital channels like Netflix, Hotstar, and so many more like these. COVID-19 halted the global film industries, shut cinema halls and auditoriums, and is also brining about a major change in how he public has now started looking at means of media as entertainment (Hall & Pasquini, 2020).

In the state of California in the USA, which is the seat of Hollywood, 16 industries were announced by the Governor as vital to the infrastructure of the stage, but nowhere did they mention the critical importance of the entertainment industry. People from television, radio, film industry were left with no back up and everyone in the cast crew and production were affected to a different degree by the pandemic (Johnson, 2021) Cinema theaters and studios were closed, which generated large revenue thus impacting the employees as well as the artists financially and psychologically (Nelson & Friedman, 2021) It was only months after the first lockdown was announced that the various sectors of industries collaborated their efforts to re-start work with due precautions.

As per the estimates of the American Alliance of Museums, museums across the USA have lost 33 billion dollars daily over the span of a pandemic. These are the very museums which have a daily footfall of hundreds of esthetes, history, and art lovers storming their corridors (Voon, 2020). People in the cinema industry right from directors, actors, producers, free lancers, faced the brunt of the pandemic as it started impacting their lives and livelihoods due to cancellations of shoots (Shoard, 2020).

5.3 EFFECTS OF COVID-19 PANDEMIC ON CREATIVE, ART, AND CULTURE INDUSTRY IN INDIA

India is the land of vast cultural diversity in art, music, dance, and craft. India boasts its 8 classical dances, hundreds of folk and tribal dances, classical music and regional music and arts and crafts from every nook and corner of the country. Along with that India is also the house of Bollywood, one of the biggest film industries in the world. But the diversity and big name fell short when the crisis hit the world in 2020. Country was shut down owing to the restrictions in place to curb the spread of the deadly virus due to which no performance, shows, concerts, shoots could be held and the entire at industry was staring blankly at the days passing by fearing for both one's health and profession.

Socio-economic policies already do not consider creative industry, especially the folk artists and craftsmen while planning and making policies. The cultural organizations also suffer due to lack of state support, post-pandemic it has further suffered from huge losses with 90% of the events being canceled (Tiwari, 2021).

Tiwari (2021) further adds that many of the artists and artisans already depend on agriculture and other means for livelihood as their artforms be it weaving, dance, craftsmen, folk singing cannot supplement their incomes as people have found new sources for means of entertainment and they are struggling to keep their artforms alive despite of the dire circumstances. The pandemic furthered them from the public eye, and their struggle seems real because in this state of panic, there is less of manual labor available for them to feed their families and there are no places open where they can perform either.

5.4 METHODS

A semi structured interview was administered to 5 artists who practice different art forms. A thematic analysis was conducted on the interview transcripts to understand how pandemic affected them emotionally, mentally, socially, and financially.

The following questions were asked to them in the interview:

- How did the pandemic affect your work?
- What were the emotions you experienced during the pandemic?
- How were you affected by the pandemic mentally?
- What were the coping strategies you used to deal with the change?
- Was there any social/financial support available?

Thematic analysis was used to identify the themes from the data. A theme taps important nuances about the data with respect to research problems and represents some level of patterned response or meaning within the data set (Braun & Clarke, 2006). Thematic analysis is a method for identifying, analyzing, and reporting patterns (themes) within data. It minimally organizes and describes your data set in (rich) detail. However, it also often goes further than this, and interprets various aspects of the research topic (Boyatzis, 1998).

This method had its own limitations. One being that the sample size was very small and hence a lot of data could not be gathered regarding the experiences of art professionals. Second that all the artists/art makers were from different art forms, as there was a selection bias on the part of the researcher. Further an in-depth review of literature was conducted to understand the past studies to address this issue.

On the basis of thematic analysis, the researcher came up with three themes – financial impact, psycho-social effects and coping strategies that highlight the experiences of subjects from the art and creative industry.

- **Artist 1:**
  Gender: Male.
  Artform: Mime/Theater.

Artist 1 has been practicing mime and theater for the past 20 years. He travels a lot for his shows around the country and also conducts workshops of both mime and theater. But when the pandemic hit, and lockdown was announced in the country there was nothing left for him to do. Overnight he lost all the offers he was awaiting and all his shows and upcoming workshops were canceled. He was succumbed to his one room flat for over three months with nothing left to do. He has never experienced so low in his life before.

Theme	Sub-Themes	Narratives
Financial impact	Left with no source of income for over 3 months.	"I had to request money from my non-artist friends and family."
Psycho-social effects	Anger towards the pandemic and government, feeling low and hopeless, losing sense of time and felt deprived of new ideas.	"I had nothing to do for 3 months but to stare at the walls of my room, sometimes I felt why am I even alive." "I am angry that why aren't there any contingency planning done for people who are freelances, I feel dejected."
Coping mechanism	Kept himself occupied with his somatic practice, regular exercise and writing which kept him connected to his work.	–

➢ **Artist 2:**
Gender: Female.
Artform: Indian Classical Dancer (Bharatnatyam).

Artist 2 is a practicing Bharatnatyam dancer and teacher for the last 15 years. She runs her school in New Delhi and travels the world to perform and teach. When COVID-19 pandemic hit, and lockdown was announced she had to close her classes and cancel her shows and workshops. It was financially draining and emotionally exhausting for her.

Themes	Sub-Themes	Narratives
Financial impact	Cash crunch, financially drained.	"It was difficult to keep the studio running with no in-flow of money as parents refused to pay the fee. My Shows were canceled, which was an added financial burden on me."
Psycho-social effects	Emotionally drained, unsure about the future, hopelessness, received social support.	"I am a very active person, suddenly being cooped up at home made me feel tied up." "My parents were very supportive and regularly motivated me to start dancing again and make the most of media and internet."
Coping mechanism	Turned to dancing for self, started online classed.	"Once I started online classes, which was difficult initially, I felt re-energized and even started doing my on *riyaaz* (practice). I started loving being at home after years." "Online teaching may not match in-personal classes, but it is a ray of hope and my students have now become accustomed to the online setup."

➢ **Artist 3:**
Gender: Male.
Artform: Musician (Tabla Player).

Artist 3 has been a Tabla player and *sangatkaar* for as long as he can remember. His father and grandfather were also table players. He usually plays at Kathak classes and whenever there are performances, and these are only sources of income for him. But with shutting down of auditoriums and cancellations of all the events he found himself hand to mouth as there was nothing left for him to do.

Themes	Sub-Themes	Narratives
Financial impact	Financial drain, no money to feed family with dwindling savings.	"Playing tabla is the only thing I know, I sangat with dancers in their classes and performances. But after lockdown was announced, it was difficult for me to survive and take care of my family. I have some table students too, but they were unable to come to the classes because of the curfew imposed. I didn't have resources to conduct classes online."
Psycho-social effects	Hopelessness, fear of future, insecurity.	"I was broken, I didn't know how to feed my family. Those three months of the first lockdown seemed never ending. I couldn't even cry in front of my family."
Coping mechanism	Help came from senior artists.	"Many senior artists and government provided some relief in form of money and resources that helped us feed ourselves."
		"I am learning to teach online so that I can be prepared for the future."

➢ **Artist 4:**
Gender: Male.
Artform: Contemporary Dancer.

Artist 4 is a Bengaluru based international contemporary artists. He trained in contemporary dance in Chicago, the USA. He has performed at major festivals across the world. Pandemic provided him a space to relax and work on new choreographies.

Themes	Sub-Themes	Narratives
Financial impact	Cash crunch, financially drained due to show cancellations.	"I run a dance studio and have enough savings, but I can only imagine freelancers who solely depend on their artform for livelihood."
Psycho-social effects	Pandemic provided an opportunity to calm down and relax, Hope for things to get better, excited to get along with the new normal.	"I have been taking classes online for international students, and when opportunity provided, I quickly switched to digital medium and shift my classes online. It was exciting and digital platform ensured there was no break in practice."
Coping mechanism	Dance as a coping strategy, did improvisation and choreography to beat the stress.	"Dance has always been my got to stress buster. I danced my stress and the situation also provided me to try out new improvisation depicting the pandemic in my performance."

- **Artist 5:**
  Gender: Male.
  Artform: Light Designer.

Artist 5 is an ardent lover of theater and joined a drama group in his college years. He has acted on stage, but what caught his interest in theater was backstage work especially light designing and so he decided to make a career out of it. He trained in light design from the USA, and he has been a light designer for 15 years and it is the only source of income he has. But things took a sharp turn once the pandemic hit.

Themes	Sub-Themes	Narratives
Financial impact	No work, difficulty paying bills due to dwindling savings during lockdown.	"I earn whenever there is a show or an event where I do light designing. But in the lockdown, I was broken financially, I had no money for my family despite my qualifications."
Psycho-social effects	Stress, hopelessness, suicidal ideation, fear.	"There were times when I cried hiding from my family because I was not able to earn during the lock down. For the first time I had second thoughts on my career choice."
Coping mechanism	Support came in the form of social and monetary support from friends and the government.	"I am grateful for all the actors and directors I have worked with who helped me in the trying time."

5.5 DISCUSSION

The creative, art, and culture industry includes performing arts like classical dance, folk dance and music, classical music, theater, puppetry, visual arts, craft work like weaving and pottery and also museum industry, fashion industry and cinema. These sectors always receive the bare minimum from the governments in terms of budget and revenue even in the normal times. Pandemic brought a whirlwind in the creative sector. These are the very sectors that contribute a lot to the tourism industry too and attract large number of tourists and visitors. As seen above, the pandemic affected those from art, culture, and creative industries financially, psycho-socially, and all of them coped with the situation in their own individual way. This chapter describes the experiences of people from art, culture, and creative industry in India, but findings corroborate with the creative industry and people working in it from other countries as well.

The creative industries' workers could not continue to perform and practice due to social distancing rules and the closure of all creative and cultural venues, and had to switch to virtual live performances instead, which was again possible for select few who could afford to and were digitally literate (Banks & O'Conner, 2020).

It is important to speak about the impact of the COVID-19 pandemic because it has been devastating for every sector across the globe and people from every industry are reciprocating the feelings. Due to COVID-19 every sector had to reorganize and rethink their service strategies to be able to survive the crisis. All the sectors right from business, education, logistics to art, culture, media, and creative industries have been affected. But at the same time, it has been a testing time to reassess the future of the various art, culture, and creative industries and to figure out future plans to adapt to a global crisis of this level. This has also been a time of innovations at world platform to bring art and culture to the homes of art enthusiasts and to be creative and think out of the box, or think in the box and fit oneself on the laptop screens and continue to perform, practice, and present one's artform or creation.

The pandemic not only affected the artists, art, and creative industry financially but it had a psycho-social impact too. On the one hand was the fear of a microorganism that had wreaked havoc in the world and how financially draining it was for the governments and the common man too. But on the other hand, the pandemic also impacted the mental health of people from all spheres of life. The main psychological impacts of the pandemic have been fear, anxiety, loneliness, depression, and suicidal ideation (Visanich & Attard, 2020). In a survey conducted in Germany brought the fore existential threat and serious mood disorders faced by those involved with the art and creative industry (Frick, Tallon, Gotthardt, Seitz, & Rakoczy, 2021). Ivcevic, Grossman, & Ranjan (2020) inspired by Barron's description of artists "occasionally crazier, yet adamantly saner" conducted a study to assess the vulnerabilities and resources of the artists during COVID-19 pandemic; they found that the people in arts with better resources had fewer psychological vulnerabilities.

The current study makes an effort to understand how financially draining the pandemic has been and the psycho-social impact of the pandemic on the art, culture, and creative industry in particular to understand how it impacted the artists, art makers-providers, creative people, media personals and those running museums, galleries, studios, theaters, and auditoriums. Further, it tries to shed light on the coping mechanism employed to handle the demands put by the crisis. In this study art makers from different fields were interviewed to understand and get a better understanding of their experience, both financial and psycho-social during the pandemic and what were the coping strategies they employed to deal with the 'new normal.'

5.5.1 FINANCIAL AND PSYCHOSOCIAL ISSUES SUFFERED BY PEOPLE IN ART, CULTURE, AND CREATIVE INDUSTRY

Art, culture, and the creative industry has been devasted by the pandemic and the effects have also shown on the mental health. According to Flore, Gaylor, & Hendry (2020) the arts industry experienced and empty schedule in the year 2020 which provided them livelihood, a purpose of life and a rhythm in life. This impacted their mental stability to a great extent. On the basis of their survey, they further add that all the artists even reciprocated each other's feelings of being insecure and depressed by the 'new-normal.'

According to Taking temperature report by the British Council and Art X, 67% of the people who find employment in the creative sector have been skeptical during the lockdown about surviving for a year if the lockdown continues. While 90% of them were sure that long term social impacts of pandemic are here to stay for a long time because COVID-19 pandemic has restricted the activity of artists, art makers and esthetes to a great extent (UNESCO, 2021).

A survey conducted by Firstpost (2021) says that when the lockdown of 2020 brought life to a grinding halt, it provided the art industry an opportunity to introspect how they can navigate the crisis and also update themselves with the ever more challenges of future. Randy (2021) in his report noted that COVID-19 pandemic had a devastating impact on the creative sector of America and the art and creative industry will progress only slowly from here onwards.

Dhar (2021) connected with various free-lance artists and their troupes and found out that it has been extremely difficult for most of them to make ends meet. Many bhajan singers and qawwals, who sing spiritual songs in Hindu temples and Muslim prayer halls found themselves hands to mouth and extremely dejected with the complete shutdown of temples, religious activities, and dargahs.

Flore, Gaylor, & Hendry (2020) have also noted in their survey that artists have suffered from hopelessness, as has been a finding of the current study too, and have felt the loss of power to win over the situation. The repeated lockdowns and shutting of all art centers has been detrimental for not only the artists but also the community at large, as one performer noted: *"These feelings have become more pronounced since the second lockdown ... feeling a strong disconnection from community (creative or otherwise) ... I feel despair over the state of the arts sector and how this will impact the rest of society."* The artists feel hopeless in the state of pandemic amidst lockdown, they have a fear of their artistic skills rusting and of course fear of survival in the industry with the upcoming young generation.

There are many crafts and art forms that in today's time depend solely on tourism to keep them thriving. During and after the lockdown there have been many travel restrictions on national and international travel and people have experienced 'travel fear' as well which has prevented them from traveling (Zheng, Luo, & Ritchie, 2021). Under these circumstances, artists such as puppeteers have hugely suffered due to lack of tourists who become a major source of their income. The puppeteers in Rajasthan have been very disappointed by the government and have struggled a lot to make ends meet because puppet shows and selling the wooden puppet is their main source of livelihood (ANI, 2021). In Odisha too puppetry is an integral part of their cultural nexus. Puppet shows find a place in all fairs, festivals, and even wedding functions. But due to the restrictions imposed by the government over the number of members attending any function, puppeteers have been impacted financially and are in fear of any more lockdown as the first wave already broke their back and morale (Senapati, 2021).

The physical art practitioners have been vulnerable to psychological effects due to separation from their spaces both physical and mental and at the same time fear of unemployment fear of pandemic have added to the distress (Stuckey, Richard, Decker, Aubertin, & Krielaars, 2021).

The pandemic also revealed the great divide in the art, culture, and creative sector in India that exists today, that between artists and creatives who have access to digital platforms and those who do not have the access and digital literacy. The urban Indian art community became accustomed after a brief pause of the first lockdown to the digital platform to share their artwork and continue their teaching and learning. Zoom, Facebook,

YouTube, and Instagram lives became a new norm for the creative *junta.* Organizations like the National Center for Performing Arts (NCPA), Royal Opera House Mumbai, the Museum of Art and Photography (MAP), Attakkalari, events such as the Jaipur Literature festival all moved to the digital platform and provided opportunities to the urban artists. But on the other hand, the folk artists and free lancers were begging for work and opportunities, especially during the first lockdown and even later (Acharekar, 2021).

The digital platform also came with its own downside. The artists who turned to the online medium to explore new avenues experienced reduced attention spans, digital fatigue, pressure of competition as everyone who had access to the internet had jumped onto some medium to keep their work alive. The competition also made individuals spend finances from their personal reserves to exist in the digital world, as if their absence from the digital stage would wipe out their artform.

While those in the urban areas had their fair share of issues in keeping their work alive, those in rural areas who have no access to internet or internet cannot help meet their needs as they need physical markets to sell their art or craft suffered from other psychosocial issues such as alcoholism and domestic violence. These issues were due to loss of or temporary absence of means of livelihood which increased frustration and anger in the artisans (Acharekar, 2021).

5.5.2 COPING STRATEGY EMPLOYED BY PEOPLE IN ART, CULTURE, AND CREATIVE INDUSTRY

There are numerous tribes and folk cultures across the world, from the indigenous Americans to aboriginals, and many central Asian tribes. Each of them has their own folk habits, music, and dances which are diverse forms of creative expression and are a panacea for a healthy life since ancient times. The COVID-19 revealed the vulnerability of the art, culture, and creative industry and magnified it for the policy makers, governments, and NGOs to look into the matter and make amends. The pandemic also unearthed the underlying issue of the social and economic rights that artists and those on creative industries must enjoy.

India in particular and the world in general is a land of diverse tangible and intangible cultural heritage. Scores of folk dances and styles of folk music find an important place in the daily lives of people. In India, the

Indian classical dances grace the stages with their geometric yet rhythmic movements. Music schools, durbars, concerts are ripe with the *ragas* and *aalaps* of classical music tradition of the country. Mesmerizing architectural marvels attract a large number of tourists every year and the museums are abuzz with esthetes. Film industry that is an employer of large number of people is always producing source of entertainment for the public. Similarly, the classical Ballet, contemporary, jazz, and other styles of dances as well as famous Broadway shows grace the stage of prestigious art centers across the world.

But all this came to a sudden halt when the first lockdown was announced owing to the fear of COVID-19. Festival celebrations and the folk music and dance that accompany them could not be seen or heard in India. The thumping of feet and the sound of *ghungroos* at dance institutes and auditoriums could not be heard. The backstage buzz before a performance was quietened. No ragas and aalaps made the air lyrical. Corridors of the museums and tourist places were abandoned. The vibrant film industry also paused all its work. Even after the first lockdown was partially lifted, COVID-19 pandemic protocols were laid down by the government to ensure the safety of the public (Bhattacharya, 2021). It was extremely necessary to develop resilience during this time in the entire creative industry to ensure it thrives when the tide passes.

In many African countries too, festivals are colorful and full of merry. People get dressed in their traditional attires, sing, and dance, and is an occasion for them to come together as a community. But owing to COVID-19 the festivals people were asked to celebrate their festival in their own home, which was not how people enjoy it (UNESCO, 2020).

Coping and developing resilience during the time of pandemic is not possible at an individual level, it will need a community support. The art and creative industry are for and by the community and thus will need the support of everyone to overcome the set back that this industry has faced.

Support came in many ways one of them being financial too. The government of Luxembourg announced financial and social assistance scheme for freelance artists and employees of the entertainment industry (UNESCO, 2020). France became the first country to help the freelancer artists and art creators followed by Britain, which rolled out a financial package to save many auditoriums, comedy clubs, music venues, and theaters. These moves were further copied by countries like Germany, South Africa, Poland, and New Zealand (Marshall, 2021).

Many organizations such as Artists Fellowship Inc., Craft Emergency Relief Fund, Artists' Charitable fund, American Guild of Musical Artists Relief fund, etc., in the USA came forward to provide relief for the art industry during and after the pandemic (Artwork Archive, n.d.). While in the United Kingdom Coronavirus Job Retention Scheme (CJRS) for artists was started to benefit those employed by the art and creative industry (Towse, Frey, & Cowen, 2020).

Coronavirus though imposed social distancing but brought about solidarity like never before. Many well to do artists in India started crowd-funding online to collect funds for artists who could not switch to the online platform overnight and depend on performances and rehearsals for their daily wages (Phukan, 2020).

David Brown, Director of Policy and research at Jamaica Creative in an interview for survey reported that during the forced culture of isolation and social distancing helped in transmission of cultural knowledge from parents to children, and also among the adults who had a new found awareness and love for their respective heritage. This became a lesson for them and also a means of coping with the pandemic (UNESCO, 2021, p. 37). This brought about a revival of various aspects of intangible heritage in various parts of the world, for example lacemaking, embroidery techniques, traditional crafts, cooking of traditional dishes which had so far been forgotten in Iran, oral storytelling traditions of China and traditional games from Italy and Palestine and weaving tradition from Ecuador were taught to the younger generation many of whom experienced catharsis using them and who are now creating opportunities in their new found knowledge (UNESCO, 2021, pp. 38, 39). Thus, it can be seen that world over the lockdown served as a means of people being isolated from their routines but returning to their roots and became a blessing in disguise as it led to a revival of many art forms by the young generation.

In many parts of the world were artists and creatives were marinating social distancing and struggling to make their ends meet, sought spiritual and emotional support from the very same living heritage that they practice (UNESCO, 2021, p. 41). The digital world opened new avenues for the creative sector to display and share the art and creativity for the non-artists who turn to these forms for their well-being and entertainment. In an interview to The Hindu, Smriti Rajgarhia who heads the Serendipity Arts festival said that *"The Internet brought to the forefront unique opportunities for artists and institutions to interact with each other in a varied manner, forging collaborations and creating newer outlets for the arts."* (Acharekar, 2021).

Instead of finding means to cope with the crisis situation, art itself became a system of support through the digital means. Videos of dance traditions were shared initially by the Islamic Republic of Iran's medical staff in the hospital, to boost morale and escape feelings of grief and sorrow, but soon became viral among the larger public, while in Kyrgyzstan traditional storytellers who had lost their means of livelihood became motivators through their story telling online to boost morale of people and to stay one in these trying times (UNESCO, 2021, p. 42).

Pandemic also brought a tough time for the global film industry as everyone was locked in and the shoot locations were shut down. But it did not stop the creatives to plunge back in the race and innovate new means of film production. In spite of being isolated they found ways of working virtually to create cinema and this change in film making is here to stay. Of course, not everyone who finds employment in the film making could be benefitted but it was a start to something (Atkinson & Kennedy, 2020).

As discussed in the previous section, many cultural organizations shifted their bases online to support, promote, and sustain creative opportunities for the artists. Kolkata cultural Center went a step further in providing opportunities for the artists who come from rural and tribal areas to perform and sell their works online. Many state governments also offered to help artists with an amount of 5,000/– to help sustain their life in the trying times (Hindustan Times, 2021). Rajasthan government and West Zone Cultural Center went a step further in coming up with a scheme 'Chief Minister's Folk Artist Incentive Scheme' to protect the self-respect of artists in the state (PTI, 2020). The Delhi government also took up the cause of creating an online database for the artists to provide the organizers with access to artists' data which can help the latter get work (Barman, 2021).

Towards the end of the year 2020, several workshops, webinars, concerts, and performances had come up on the digital platform which had a two way benefit. The artists as well as the audience, learners, and art connoisseurs benefitted from the creative activities. Arts are healing in any form be it painting, crafts, dance, music or dramatics, keeping this in mind Attakkalari center of arts in Bengaluru also took up the cause to bring to the masses free of cost digital lessons. These movement and music sessions created an online space for catharsis and relaxation.

In an interview to The Hindu, Jayachandran Palazhy, director of Attakkalari said *"The two-dimensional screen, definitely, is a limitation. To share a space with others, sweating it out together is always better. The online space is, nevertheless, a wonderful domain that lets us reach out to more people. We might not entirely migrate to this space. But we will continue to use it more often, even after things are back to normal"* (Sudevan, 2020).

Many artists who have access to regular flow of internet started their classes and performances online. Of course, adjustments had to be made with respect to space, quality of internet on both ends of sender and receiver and online fatigue had to be dealt with, but people didn't back down and they accepted this as an opportunity only to get back stronger.

In this tying time UNESCO launched ResiliArt movement which opened the stage for global debates by eminent art makers and creatives in support of the art world and to show resilience with the aim to bounce back (Un News, 2020).

5.6 CONCLUSION

The year 2020 was the year of artistic resilience (Ganguly, 2020). Artists, esthetes, creatives, they all stood up for each other in the difficult time to allow each other space to grow, expand, create, and re-build oneself. Performing, creating, choreographing, installing, and all the terms that are associated with the creative industry can never be as best as they are in the physical spaces. But the digital platform provided an opportunity to transform what looked like the end of indigenous art forms, performances, and creative spaces. For sustained income avenues, and to increase its benefits the sector has to innovate by finding stronger connections across related areas like education and health (Padampriya & Verma, 2020) and also to increase the digital maturity to enhance productivity and efficiency (Fletcher & Griffiths, 2020). Socio-ecological programs and resilience promoting programs can be developed using the arts and creative resources to help the creative fraternity as well as the ones who reap the benefits of the art and creative activities (Stuckey, Richard, Decker, Aubertin, & Krielaars, 2021).

Currently there is dearth of researches on artists in general on psycho-social impact of COVID-19 and holistic health interventions, but the need for the same has been realized by experts in the field to promote coping interventions and resilience training for the art and creative sector to face future crisis and to overcome the current one (Ménard & Hallé, 2014; Filho et al., 2016; Burtt & Lavers, 2017; Stuckey et al., 2021).

Art has the quality to build solidarity, to inspire, pacify, and allow for thoughts and emotions to be shared (UN News, 2020). The UN News further reported (2020) that amidst the COVID-19 pandemic's first wave that had spread like wildfire across the world, UNESCO celebrated World Art Day on April 15, 2020, which served as a reminder that arts have a power to connect even when everything seems to be falling apart.

Art and culture are at the heart of our life and social existence. One thing that art teaches in the times of crisis is resilience. Artistic activities involve creativity, and creativity means bringing innovation, changing the norm, bringing something new on the plate. As art enthusiasts, esthetes, artists, creatives we must resolve to support the creative industry through COVID-19, build global solidarity, and build emergency support for the entire industry to be able to face any such crisis in future, and to create a sustainable platform to showcase and exhibit all forms of art culture and creativity be it in the form of dance, drama, music, museums, and even the travel industry that is directly impacted by any crisis in the creative sector (World Economic Forum, 2020).

Thus, despite of the hardships that the art, culture, and creative industry has suffered over the course of the pandemic, it will always rise like a phoenix and bounce back and rebuild itself again as it is a mirror to the society and its changes, and both shape each other for the better.

KEYWORDS

- **art**
- **artists**
- **coronavirus job retention scheme**
- **COVID-19 pandemic**
- **culture**
- **performing arts**
- **psychological effect**

REFERENCES

(2021). *Impact of the COVID-19 Pandemic on the Arts and Cultural Heritage*. Available at https://en.wikipedia.org/wiki/Impact_of_the_COVID-19_pandemic_on_the_arts_and_cultural_heritage#cite_note-28 (accessed on 7 February 2023).

Acharekar, J., (2021). *Surviving the Pandemic Has Forced Artists to Make a Virtue of Being Online*. The Hindu. Retrieved from: https://www.Survivingthepandemichasforcedartiststomakeavirtueofbeingonline (accessed on 7 February 2023).

American Library Association, (2020). *ALA Executive Board Recommends Closing Libraries to Public*.

ANI, (2021). *Rajasthan CM Announces Financial Assistance of ₹5000/– to Artists in State*. Hindustan Times. Retrieved from: https://www.hindustantimes.com/ lifestyle/art-culture/rajasthan-cm-announces-financial-assistance-of-rs-5000-to-artists-in-state-101623382075372.html (accessed on 7 February 2023).

Atkinson, S., & Kennedy, H. W., (2020). *This is How the Film Industry is Fighting Lockdown*. World Economic Forum. https://www.weforum.org/agenda/2020/05/this-is-how-the-film-industry-is-fighting-lockdown (accessed on 7 February 2023).

Banks, M., & O'Connor, J., (2021). Art and culture in the viral emergency. *Culture Trends, 30*(1), 1, 2. 10.1080/09548963.2021.1882063.

Barman, S. R., (2021). *Artists Hit by COVID and Lockdowns, Delhi Govt Plans Database to Get Them Work*. The Indian Express. Retrieved from: https://indianexpress.com/article/cities/delhi/artists-hit-by-covid-and-lockdowns-govt-plans-database-to-get-them-work-7397425/ (accessed on 7 February 2023).

Bhattacharya, S., (2021). *The Show Must Go On: Guru-Shishya Parampara in Post Pandemic World. Sahapedia*. Retrieved from: https://www.sahapedia.org/show-must-go-guru-shishya-parampara-post-pandemic-world (accessed on 7 February 2023).

Boseley, M., & Convery, S., (2020). *Carriageworks Goes Into Voluntary Administration Citing 'Irreparable Loss Of Income' Due To Coronavirus*. The Guardian.

Boyatzis, R. E., (1998). *Transforming Qualitative Information: Thematic Analysis and Code Development*. Thousand Oaks, CA: Sage.

Braun, V., & Clarke, V., (2006). Using thematic analysis in psychology. *Qualitative Research in Psychology*. Retrieved from: https://www.researchgate.net/publication/235356393_Using_thematic_analysis_in_psychology (accessed on 7 February 2023).

British Council, (2020). *COVID-19's Impact on the Arts Sector in China*. https://www.britishcouncil.org/research-policy-insight/insight-articles/impact-covid-19-arts-sector-china (accessed on 7 February 2023).

Dhar, S., (2021). *Reduced Govt Spending, Pandemic Make Survival Tough for Artistes*. The Times of India. Retrieved from: https://timesofindia.indiatimes.com/india/report-reduced-govt-spending-pandemic-make-survival-tough-for-artistes/articleshow/82176874.cms (accessed on 7 February 2023).

Filho, E., Aubertin, P., & Petiot, B., (2016). The making of expert performers at cirque du soleil and the national circus school: A performance enhancement outlook. *J. Sport Psychol. Action, 7*, 68–79. doi: 10.1080/21520704.2016.1138266.

Firstpost, (2021). *A Year Since India's COVID-19 Lockdown, an Overview of How Artists and Cultural Institutions Navigated a Crisis*. Retrieved from: https://www.firstpost.com/art-and-culture/a-year-since-indias-covid-19-lockdown-an-overview-of-how-artists-and-cultural-institutions-navigated-a-crisis-9458211.html (accessed on 7 February 2023).

Fletcher, G., & Griffiths, M., (2020). Digital transformation during a lockdown. *International Journal of Information Management, 55*. 10.1016/j.ijinfomgt.2020.102185

Florida, R., & Seman, M., (2020). *Lost Art: Measuring COVID-19's Devastating Impact on America's Creative Economy*. Brookings. https://www.brookings.edu/research/lost-art-measuring-covid-19s-devastating-impact-on-americas-creative-economy/ (accessed on 7 February 2023).

Frick, U., Tallon, M., Gotthardt, K., Seitz, M., & Rakoczy, K., (2021). Cultural withdrawal during COVID-19 lockdown: Impact in a sample of 828 artists and recipients of highbrow culture in Germany. *Psychology of Aesthetics, Creativity, and the Arts*. https://doi.org/10.1037/aca0000389.

Ganguly, N., (2020). *2020 Will be Remembered as the Year of Artistic Resilience*. The Hindu. Retrieved from: https://www.thehindu.com/entertainment/art/2020-will-be-remembered-as-year-of-artistic-resilience/article33411547.ece (accessed on 7 February 2023).

Ghoshal, S., (2020). *How COVID 19 has Darkened India's Cultural Season*. Livemint. https://lifestyle.livemint.com/how-to-lounge/art-culture/how-covid-19-has-darkened-india-s-cultural-season-111604032549060.html (accessed on 7 February 2023).

Hal, S. B., & Pasquini, S., (2020). *Can There be a Fairy-Tale Ending for Hollywood After COVID-19?* World Economic Forum. https://www.weforum.org/agenda/2020/07/impact-coronavirus-covid-19-hollywood-global-film-industry-movie-theaters/ (accessed on 7 February 2023).

Ivcevic, Z., Grossman, E., & Ranjan, A., (2020). Patterns of psychological vulnerabilities and resources in artists and nonartists. *Psychology of Aesthetics, Creativity, and the Arts*. Advance online publication. https://doi.org/10.1037/aca0000309.

Jankiraman, P., & Verma, M., (2020). *The Arts Sustained Many Through COVID, but it Leaves Behind Formidable Challenges for the Artists*. Indian Express. https://indianexpress.com/article/opinion/columns/street-performers-in-india-covid-pandemic-creative-economy-7120344/ (accessed on 7 February 2023).

Johnson, M., (2021). Hollywood survival strategies in the post-COVID-19 era. *Humanit. Soc. Sci. Commun.*, 8. https://doi.org/10.1057/s41599-021-00776-z.

Kenney, N., (2020). *Led by Metropolitan Museum of Art, a Flurry of US Museums Say They Are Shutting Down Because of Coronavirus*. The art newspaper. Retrieved from: https://www.theartnewspaper.com/2020/03/12/led-by-metropolitan-museum-of-art-a-flurry-of-us-museums-say-they-are-shutting-down-because-of-coronavirus (accessed on 7 February 2023).

Khlystova, O., Kalyuzhnova, Y., & Belitski, M., (2021). The impact of the COVID-19 pandemic on the creative industries: A literature review and future research agenda. *Journal of Business Research, 139*, 1192–1210. https://doi.org/10.1016/j.jbusres.2021.09.062.

Li, F., (2020). The digital transformation of business models in the creative industries: A holistic framework and emerging trends. *Technovation, 92, 93*(102–112). doi: 10.1016/j.technovation.2017.12.004.

Liscia, V. D., (2020). *A Daily Report on How COVID-19 Is Impacting the Art World.* Hyperallergic. https://hyperallergic.com/546913/coronavirus-daily-report/ (accessed on 7 February 2023).

Marshall, A., (2021). *How 8 Countries Have Tried to Keep Artists Afloat.* New York Times. https://www.nytimes.com/2021/01/13/arts/coronavirus-pandemic-arts-support.html (accessed on 7 February 2023).

Ménard, J. F., & Hallé, M., (2014). Circus also needs performance psychology: Facts and realities of consulting at cirque du soleil. In: Cremades, J. G., & Tashman, L. S., (eds.), *Becoming a Sport Exercise and Performance Psychology Professional: A Global Perspective* (127–134). New York, NY: Psychology Press.

Nelson, E., & Friedman, G., (2020). *The Owner of Regal Cinemas is Closing its U.S. Theaters, with 40,000 Jobs at Stake.* New York Times. https://www.nytimes.com/live/2020/10/05/business/stock-market-today-coronavirus#regal-cinemas-closing (accessed on 7 February 2023).

OECD, (2020). *Culture Shock: COVID 19 and the Cultural and Creative Sectors.* https://www.oecd.org/coronavirus/policy-responses/culture-shock-covid-19-and-the-cultural-and-creative-sectors-08da9e0e/ (accessed on 7 February 2023).

Phukan, V., (2020). *How to Support the Arts During Pandemic.* The Hindu. https://www.thehindu.com/entertainment/theater/how-to-support-the-arts-during-a-pandemic/article31312549.ece (accessed on 7 February 2023).

Pitchfork, (2020). *Coronavirus: Updated List of Tours and Festivals Cancelled or Postponed Due to COVID-19.* https://pitchfork.com/news/coronavirus-updated-list-of-tours-and-festivals-canceled-or-postponed-due-to-covid-19 (accessed on 7 February 2023).

Press Trust of India, (2020). In: Bharat, R., (ed.), *Work Opportunities Should Be Given to Artists Hit by Lockdown: Rajasthan Governor.* Retrieved from: https://www.republicworld.com/india-news/politics/work-opportunities-should-be-given-to-artists-hit-by-lockdown-rajasthan-governor.html (accessed on 7 February 2023).

Radermecker, A. S. V., (2021). Art and culture in the COVID-19 era: For a consumer-oriented approach. *SN Business and Economics, 1*(4). https://doi.org/10.1007/s43546-020-00003-y.

Randy, C., (2021). *COVID-19's Impact on the Arts.* Americans for the Arts. https://www.americansforthearts.org/node/103614 (accessed on 7 February 2023).

Rieger, D., Reinecke, L., Frischlich, L., & Bente, G., (2014). Media entertainment and well-being—Linking hedonic and eudaimonic entertainment experience to media-induced recovery and vitality. *Journal of Communications, 64*(3). doi: 10.1111/jcom.12097.

Rodríguez-Gulías, M., Fernández-López, S., & Rodeiro-Pazos, D., (2020). Innovation in cultural and creative industries firms with an academic origin (CCI-USOs): The role of regional context. *Technovation, 92, 93,* 10.044. doi: 10.1016/j.technovation.2018.06.007.

Romine, T., Levenson, E., & Morgado, J., (2020). *Broadway Theaters to Suspend all Performances Because of Coronavirus.* CNN. Broadway shows suspended in response to coronavirus fears – CNN (archive.org).

Senapati, A., (2021). *COVID-19 Second Wave: Livelihoods of Odisha's Puppeteers Hanging by a Thread.* Down to Earth. https://www.downtoearth.org.in/news/economy/covid-19-second-wave-livelihoods-of-odisha-s-puppeteers-hanging-by-a-thread-77243 (accessed on 7 February 2023).

Shoard, C., (2020). *'Over one Hour Everything was Cancelled' – How Coronavirus Devastated the Film Industry*. Guardian. https://www.theguardian.com/film/2020/mar/20/over-one-hour-everything-cancelled-coronavirus-impact-film (accessed on 7 February 2023).

Stuckey, M., Richard, V., Decker, A., Aubertin, P., & Kriellaars, D., (2021). Supporting holistic well-being for performing artists during the COVID-19 pandemic and recovery: Study protocol. *Frontiers in Psychology*. https://doi.org/10.3389/fpsyg.2021.577882

Sudevan, P., (2020). *Bengaluru Movement Arts Center Attakkalari is Conducting Online Programs for Children and Adults*. The Hindu.

The Conversation, (2020). *Parts of Life Will Be Damaged Forever' — Arts Workers Describe the Pandemic's Impact on Their Mental Health*. Retrieved from: https://theconversation.com/parts-of-life-will-be-damaged-forever-arts-workers-describe-the-pandemics-impact-on-their-mental-health-147252 (accessed on 7 February 2023).

The Conversation, (2020). *Research Shows Pandemic's Toll on Arts and Creative Sector Workers' Mental Health*. Research shows pandemic's toll on arts and creative sector workers' mental health – RMIT University.

Tiwari, P., (2020). *Protecting Artists and the Arts*. The Hindu. Protecting artists and the arts – The Hindu.

Towse, R., Frey, B. S., & Cowen, T., (2020). *What Financial Support is Needed for Artists and the Arts in the Pandemic?* Economics Observatory. https://www.economicsobservatory.com/what-financial-support-needed-artists-and-arts-pandemic (accessed on 7 February 2023).

UN News, (2020). *Even During COVID-19, Art 'Brings us Closer Together Than Ever' – UN Cultural Agency*. https://news.un.org/en/story/2020/04/1061802 (accessed on 7 February 2023).

UNESCO, (2020). *COVID-19 Impact on Unique Girls' Festival in Ethiopia*. https://ich.unesco.org/en/living-heritage-experience-and-covid-19-pandemic-01124?id=00308 (accessed on 7 February 2023).

UNESCO, (2020). *Social Assistance Scheme for Luxembourg's Freelance Artists*. https://en.unesco.org/creativity/covid-19/social-assistance-scheme-luxembourgs-freelance (accessed on 7 February 2023).

UNESCO, (2021). *COVID 19 Impact Report on India*. https://en.unesco.org/creativity/covid-19/covid-19-impact-report-india (accessed on 7 February 2023).

UNESCO, (2021). *Living Heritage in the Face of COVID-19*. https://unesdoc.unesco.org/in/documentViewer.xhtml?v=2.1.196&id=p::usmarcdef_0000377671&file=/in/rest/annotationSVC/DownloadWatermarkedAttachment/attach_import_7c0cab02-450d-40ad-a8e0-273bfc689842%3F_%3D377671eng.pdf&locale=en&multi=true&ark=/ark:/48223/pf0000377671/PDF/377671eng.pdf#%5B%7B%22num%22%3A155%2C%22gen%22%3A0%7D%2C%7B%22name%22%3A%22XYZ%22%7D%2C0%2C842%2Cnull%5D (accessed on 7 February 2023).

Visanich, V., & Attard, T., (2020). Performing precarity in times of uncertainty. *COVID-19 Volume II: Social Consequences and Cultural Adaptations*. https://www.google.co.in/books/edition/COVID_19/8qoLEAAAQBAJ?hl=en&gbpv=0&kptab=overview (accessed on 7 February 2023).

Voon, C., (2021). *How Three Very Different Museums Are Dealing with the COVID-19 Crisis.* Artsy. https://www.artsy.net/article/artsy-editorial-three-museums-dealing-covid-19-crisis (accessed on 7 February 2023).

Watson, M., & Harmon, S., (2020). *Coronavirus: Australia's Performing Arts Industry 'Could be Brought to the Brink.* The Guardian. https://www.theguardian.com/artanddesign/2020/mar/14/coronavirus-australias-performing-arts-industry-could-be-brought-to-the-brink (accessed on 7 February 2023).

World Economic Forum, (2020). A*rt and Culture Global Solidarity Network.* https://www.ala.org/news/press-releases/2020/03/ala-executive-board-recommends-closing-libraries-public (accessed on 7 February 2023).

Zheng, D., Luo, Q., & Ritchie, B. W., (2021). Afraid to travel after COVID-19? Self-protection, coping and resilience against pandemic 'travel fear.' *Tourism Management,* 83. https://doi.org/10.1016/j.tourman.2020.104261.

CHAPTER 6

Is Resilience a Predictor of Psychological Well-Being Among Employees Amid Lockdown? A Literature Review-Based Study

MONIKA GWALANI

Assistant Professor, Amity University, Jaipur, Rajasthan, India

ABSTRACT

Calamities, whether natural or human, are ruinous events that cause individual stress, disrupt community routines and dynamics, deteriorating infrastructure and businesses, and result in economic crises for a considerable time. World Health Organization (2020) declared COVID-19 as a pandemic which in an unprecedented way has affected individuals, organizations, communities, and nations. In order to combat with COVID-19, major Governments have declared lockdown in their nations, which forced many organizations and workers to work through virtual platform. The radical shift in work pattern among e-workers have caused to face occupational stress, financial insecurity, ineffective work-life balance, social, and professional isolation, poor employee engagement, low production rate and so on. The current chapter aims to investigate whether resilience proves to be a predictor of psychological well-being among e-workers amid pandemics. By reviewing conceptual and empirical studies in global context, the current study recommends that in order to foster psychological well-being, among the e-workers, there is need to adapt change and propensity to bounce back from hazard which can be

Reshaping the Business World Post-COVID-19: Management Strategies for Sustainable Behavior Change. Arvind K. Birdie & Ruchi Joshi, (Eds.)

done by collaborative efforts of the employees and companies. Employees can attain effective psychological well-being by sustaining day to day activities, develop self-efficacy, practice yoga and meditation. Organization can as well contribute, in this context, by practicing flexible working hours, organizing tele-counseling sessions through psychologists. Such intervention strategies would not only prove beneficial to the employees and their psychological well-being but will also help the organizations to flourish.

6.1 INTRODUCTION

6.1.1 COVID-19 AND ITS GLOBAL IMPACT AMONG EMPLOYEES

Calamities, natural or human-made, have always had severe physical, psychological, social, and economic negative impact on individuals (Neria et al., 2008). Novel Coronavirus Disease (COVID-19), a severe respiratory illness, broke out in China (Wuhan) in December 2019 with a conglomeration of cases with pneumonia of unknown cause (Zhu et al., 2020), which in a very short period of time not only attacked the Chinese population but has also affected the individuals from different countries of the world. Being a novel virus, which is highly infectious in nature and dispersed among people of different countries, there were no proper medical treatments available. On 11th March 2020, World Health Organization announced it as 'pandemic' caused by severe acute respiratory syndrome corona virus (SARS-CoV-2 virus). The virus spread among people through an infected person's mouth or nose in small liquid particles when they cough, sneeze, speak, sing, or breathe. These particles range from larger respiratory droplets to smaller aerosols. In an attempt to curb the plight, WHO declared some guidelines such as practice respiratory etiquette, stay home, self-isolate, avoid social gathering, stay at least 1 meter apart from others, wash hands regularly with soap and water, and wear a properly fitted mask when physical distancing is not possible. The government of affected countries followed the guidelines and imposed lockdown in their respective countries. Schools, colleges, shopping malls, entertainment, and leisure facilities, as well as ban on all public and private events and gatherings, were closed. With the promulgation of lockdown, there emerged a need to carry day to day activities through online platform. This radical transition from physical to virtual mode had tremendous negative physical, psychological, social, and financial impacts in multiple countries worldwide (Montenovo et al., 2020).

On one end, the COVID-19 pandemic generated threats including fear of losing one's own and life of near and dear ones, absence of proper medical treatment and lack of availability of vaccination, media reports related to increase in number of cases, death rate, and overcrowded hospitals (Pakenham et al., 2020), restricted mobility, and social isolation. On the other hand, the governmental imposed lockdown, due to COVID-19 pandemic, was an extra organizational stressor along with chronic workplace stressors such as role conflict, meeting deadlines in a short span of time, cutthroat competition, long hours, and heavy workload and so on. Resultantly, many of the employees of COVID affected nations noted drastic change in their daily routine. Employees who earlier used to drive to their work and had effective social life outside their home found themselves in a pandemic enforced work from home (WFH) situation. Employees from different sectors had different consequences of COVID-19 lockdown. On one end, employees working in health sector and other essentials such as supermarket were having heavy workload, on the other hand, there were some workers who were laid off or furloughed from their job due to lockdown guideline, yet there were few other employees who were working in such companies or organizations which adapted new working techniques known as smart working or e-working or work from/at home (Giorgi et al., 2020). Employees amid pandemic were working and taking care of their family hand in hand, home schooling, having limited amount of social support which is a core protective factor against psychological health issues (Halbesleben, 2006). In the absence of prior experience of WFH, many workers were having strenuousness experience in managing workspace at their home, acquiring new way to interact with the colleagues, handling with work isolation, circumference personal and work life (Ozcelik, & Barsade, 2018; Shimazu et al., 2020; Cho, 2020). Inadequate support from co-workers and poor resources to meet these challenges, due to pandemic enforced remote working affected the employees' psychological well-being (Brooks et al., 2020) which in turn negatively affected their work performance (Ozcelik & Barsade, 2018).

6.2 PSYCHOLOGICAL WELL-BEING

Psychological well-being was first recognized by Jahoda and basically presented these concepts as clinical perspectives which are a state of mental health (Usman, 2017). Wright (2010) defined psychological well-being as a subjective and worldwide experience that a person is meeting a maximal positive and generally minimal negative emotions or feelings. Psychological well-being is the potentiality of a person to perceive contentment and work effectively regardless of negative or distressing feelings which are typically part of life (Huppert, 2009). Chaturvedula and Joseph in the year 2007 defined psychological well-being as an individual's judgment or assessment of his or her life – either regarding life fulfillment (intellectual assessments) or influence (emotional reactions) which is additionally partitioned into pleasant effect (positive emotions) and unpleasant effect (negative emotions). Moe (2012) considered psychological well-being as the foundation stone of emotional wellness. Psychological well-being is simply defined as ultimately about individual satisfaction—feeling good and living safely and healthily. Psychological well-being is composed of all the short-term and long-term mental functioning and positive well-being (e.g., positive affect, self-confidence, and morale) and negative well-being (e.g., depression, discouragement, anxiety) (O'Driscoll & Brough, 2003). In 1989, Ryff proposed the multi-dimensional model of psychological well-being reflecting an individual's capacity to deal effectively with existential challenges of life. Following discussed are the dimensions of psychological well-being:

1. **Autonomy:** These endeavors individuality in social environment. It is a propensity of a person to take independent decisions in life in spite of societal pressure.
2. **Environmental Mastery/Control:** This dimension is defined as an individual's proficiency in coping with stressful events of life and grabbing the opportunities to meet personal needs.
3. **Personal Growth:** The dimension of personal growth reflects an individual's self-awareness regarding one's strengths and virtues as well as openness to experience and constantly improving to acquire skills.
4. **Positive Relations with Others:** This dimension is related to practicing empathy and caring attitude towards others, understanding the give and take of human relationships, affection, and

intimacy, and developing and managing healthy and trustworthy relationship with others.

5. **Purpose in Life:** The dimension indicates setting goal, target or possessing a sense of direction in life, it also include having meaningful past and present lives.
6. **Self-Acceptance:** Ryff (1989) defined this dimension as holding a positive attitude towards oneself, having awareness about one strengths and weaknesses and positive approach towards the past.

In the occupational sector, psychological well-being has been recognized as noteworthy in recent decades due to the radical shift from physical to psychological wellsprings of strain at work. Deriving out of different outlooks, optimal level of pressure at work is considered as psychologically healthy. As it opens the doors for an employee to experience a sense of challenge and achievement, which are important for a high level of psychological well-being. But when such work pressures are experienced frequently and excessively, then instead boosting the psychological well-being, it gets deteriorated. Well-being has turned into a dominant point in research as numerous authors found it as indispensable to workers' global well-being and growth since it embodies an all-encompassing outlook of individual health (Schoeman, 2012). Through research, it has been noted that an organization's productivity and efficiency depends upon employees' satisfaction, happiness, and good well-being (Avey, Luthans, Smith, & Palmer, 2010). Psychological well-being plays a vital role in an individual's feeling in day to day life and these feelings may vary from positive feeling such as happiness, satisfaction to negative feelings such as depression and dissatisfaction. Psychological well-being reflects how individuals evaluate their lives. These evaluations are in the form of cognition and affect. The cognitive evaluation is done through when a person gives conscious evaluative judgment about one's satisfaction with life as a whole. The affective or emotive evaluation is done through emotions and feelings which a person experience in reaction to their different life events. The more an employee experiences healthier psychological well-being, the better will be their organization's growth and profit. Conversely, staff members experiencing poor or deteriorated psychological well-being will result in absenteeism, reduced production and increased costs. In many developed countries, 35%–45% of absenteeism from work is due to mental health problems (WHO, 2003). Employees' mental health is one of the important components of occupational health. The National Institute of Safety and Health (NIOSH) recognize psychological disorders as one of the 10 pre-eminent work-related ailments and sufferings (Miller, 1984). Shagvaliyeva & Yazdanifard (2014) expressed that employees' well-being (physical, mental, and personal) may alter their work efficiency and productivity. Employees working remotely are experiencing a devastating effect on their emotional and mental health. Working for longer period of time on virtual platform, attending virtual meetings, lack of physical contact with colleagues or supervisor, digital inequalities, taking care of children and elderly family member along with online work are some of the major factors which is impacting their psychological well-being, which in turn hampering their work performance, job motivation, and finally leading to economic loss of the company.

In order to overcome the stressors, such as pandemic, there is a need to adapt certain such positive personality characteristics or traits which can help in attaining healthy and better psychological well-being of the employees so that they can work productively and proficiently and can prove to be a beneficial and resourceful asset not only for the company or organization where they are working but to the nation as well. Following are some of the major positive personality traits which may influence the psychological well-being of an employee:

1. **Grit:** A positive personality trait, is defined by Angela Duckworth (2007) as a combination of passion and perseverance for a meaningful and long-term goal. Those who possess such traits are persistent towards their target and work passionately despite of being confronted by some major difficulty or obstacle. Such individuals work diligently in the absence of any sort of positive reinforcement. Employees who secure higher score on grit in the workspace are tend to perform better and survive in their field for a longer period of time (Duckworth, Peterson, Matthews, & Kelly, 2007; Kelly, Matthews, & Bartone, 2014; Duckworth, Quinn, & Seligman, 2009). Such sorts of employees not only do their work honestly and efficiently but also manifest positive behavior at the workplace.
2. **Optimism:** As a positive psychology construct, is a set of beliefs or traits which help an individual to focus on the positive events or experiences of life instead of negative one. Buchanan & Seligman (1995) have explained optimism in the context of explanatory style, which is based on attribution theory. According to this theory,

an individual with an optimistic explanatory style will attribute favorable events to personal, permanent, and pervasive causes, and unfavorable events to external, temporary, and situation-specific causes. Contrarily, individuals with pessimistic explanatory style implies externalizing positive events and attributing them to temporary, situation-specific causes, while internalizing negative events and attributing them to permanent and pervasive causes. Seligman (1998) has also accentuated that optimism is a learnable state the can be developed over time and can have an impact in diverse life domain such as education, work, athletic performance, politics, and health.

3. **Hope:** It is defined as "a positive goal-oriented condition that is based on an interactively extracted sense of successful: (i) agency (goal-directed energy); and (ii) pathways (planning to meet goals)" (Snyder, Irving, & Anderson, 1991, p. 287). Organizational hope is a vital for studying and strengthening organizations. Based on past researches it can be said that hope is related to job outcomes such as job performance, turnover intentions, and job satisfaction clearly demonstrates that hope, in fact, impacts job outcomes (e.g., Avey, Luthans, & Jensen, 2009; Peterson & Byron, 2008; Yavas, Babakus, & Karatepe, 2013).
4. **Self-Efficacy:** It is regarded as an individual's belief in one's own potentialities. It is an individual's self-confidence in one's own skills, aptitudes, and proficiencies in achieving the set goals or targets and producing desired outcomes. High efficacious employees consider crises and dilemmas as challenges. They are highly committed towards the work or task they carry out and invest a good amount of time and energy into their work activities. They also plan strategically to deal with problems, recover easily from predicaments, feel they can control the majority of stressors and are less vulnerable to stress and depression (Bandura, 2008).
5. **Life Satisfaction:** Locke, in the year 1976, defined life satisfaction as a pleasurable or positive state resulting from the evaluation of one's work experience and life related activities. Personnel who have higher level of life satisfaction are tend to be more committed toward their organization, more productive, less likely to take leave and have satisfaction with their lives in comparison to those who have poorer life satisfaction (Lease, 1998).

6. **Coping:** Another positive personality trait is coping. Lazarus (1993, 1999) defined as a cognitive and behavioral approach that can be used by an individual to deal with tedious time and stressful situations, whether it is internal and/or external one. Working in today's fast-moving and erratic environment employees are constantly under pressure to acquire and adjust their expertise as per the demand of the organization or company. Constantly encountering stressful events at workplace may not only deteriorate the physical and psychological health of the employees, but it also affects the work production and turnover of the company (Tang, Kim, & Tang, 2000).
7. **Resilience:** A relative stable personality trait (Kirkwood, Bond, May, & McKeith, 2010), in the capacity of an individual to overcome, steers through and bounces back from adversities (Ong, Bergeman, Bisconti, & Wallace, 2006). The APA (American Psychological Association) defines resilience as the process of adapting well at the time of adversity, trauma, tragedy, threat, or even significant source of stress such as family and relationship issues, workplace or financial stressor, major medical problems. In very simple words, resilience is 'bouncing back' from a tough situation. Individuals possessing high resilience will be able to develop and maintain positive social relationships, manage their behavior and maintain a positive view of self and their outlook of lives (Wright, Masten, & Narayan, 2013).

In order to control the situation amid pandemic, the government of different nations have imposed different kind of lockdown, quarantine, physical/social distancing which drastically influenced key life aspects impacting on personal mobility (limited activities due to home quarantine), interpersonal relationship (e.g., restricted face-to-face interaction), educational/occupational activities (change in working and daily activities of workers and students). Amid such a crisis, a firm's resilience is one of the important aspects which may alter the continuity and recovery of business.

The current chapter is going to focus on how resilience can help in determining the psychological well-being of the employees working amid pandemic era. There are several psychological attributes such as work motivation, job satisfaction, work-family balance, and job crafting, etc., which were deteriorated due to remote working and caused to them to experience poor psychological well-being. The chapter is also going to help us in understanding how resilience an employee can handle calamity like COVID-19 and can adapt and adjust to the new normal. Following are the some of the psychological attributes which were compromised amid pandemic:

1. **Work Motivation and Resilience:** In any organization or company, employee is an important asset as they are the one who make rigorous effort to achieve company's goal and mission or in other words companies' growth and success depends on employees that are working. The more efficient employees a company is having the more they will progress and flourish. Hence, employee's participation and commitment play a prominent role in the success of a company. Work motivation is one of the most important element which determines an organization's productivity and efficiency. It is a pushing factor that compels a person to make efforts to attain desired goals. Work motivation is defined as the enthusiasm, energy level, commitment, and the amount of creativity that an employee brings to the organization on a daily basis. More recently, Robbins (2005) defines work motivation as "the willingness to exert high levels of effort towards organizational goals, conditioned by the effort's ability to satisfy some individual needs." Under apprehensive situations at work, it becomes a challenge for employees to attain their full potential, but this can be achieved by motivating them. Amid COVID-19 era, remote working has been the only option that the organizations are considering to undertake their businesses. Through research, it has been noted that due to the transformation in the work landscape, employees are experiencing work dissatisfaction and demotivation. The reason behind this decrease in level of work motivation are as follow:

   i. **Inadequate Social Life:** Human is a social animal, and in the absence of physical contact with other humans is affecting their physical and mental health. Due to governmental imposed lockdown, employees though are doing their job and earning money, but that is not sufficient. The meetings and discussions on virtual platforms like Google Meet, Teams, and Zoom, etc., are not satisfactory and may cause communication errors. Under such circumstance, it sometimes becomes a challenging task for employee to undertake the assigned task. Poor social contact in the long run may hamper the level of work motivation and psychological well-being as well.

ii. **Paucity of Technical Knowledge:** Amid pandemic era, organizations, and companies have to accept the governmental imposed lockdown under which they could not have their employees working physically in their company. So they adopted the 'new normal' which is e-work or telecommuting or virtual work. Employees who were having proficiency in technology and other related things were being able to carry out their task, but then there were some other employees as well who were not very techno-friendly. Such employees despite of having good knowledge and work experience about their job are unable to excel which may cause the feeling of sadness and frustration, which may further impact their work motivation.

iii. **Distraction and Disturbance from Family Members:** To reduce the transmission of COVID-19, the regional and national governments have imposed non-pharmaceutical guidelines which include school and university closures, recommendations for working from home, gym, cinema hall, shopping mall and other entertainment places were shut and bans on large gatherings. The presence of family members while attending meeting or doing the work assigned by managers may cause distraction in the long run such sort of disturbance may hamper employees' motivation.

To boost employees' motivation there is need to adapt resilience, that is, capacity to bounce back from negative situation and handle the negative emotions. Research conducted in this regard has noted that one of the key factors which determine an employee's work motivation is resilience (Martin, 2005). Employees with resilience, majorly, do not react strongly to tough situations; they would instead react calmly and experience positive emotions. Such individuals have the potential to realize the intensity of work, capacity to derive meaning from their work, and handle the situation firmly.

2. **Work Satisfaction and Resilience:** Job satisfaction refers to an individual's positive or negative attitude towards the kind of job they pursue (Spector, 1997) or the extent to which an individual perceives that his/her needs are fulfilled by the work (Evans, 1997). It may vary from extreme satisfaction to extreme dissatisfaction. The factors which may alter an employee's level of job satisfaction are both personal and institutional. The personal factors include age, gender, education (Lu et al., 2016; Aziri, 2011) on the other hand the institutional factors encompass type of occupation, work load, salary, and reward, work environment, job security, adequacy of resources, appreciation, interpersonal relationship (Lu et al., 2016; Temesgen, Aycheh, & Leshargie, 2018; Aziri, 2011). Researches done on job satisfaction have proved that level of job satisfaction determines employees' work performance, organizational commitment, absenteeism, retention, and turnover rates (Ali et al., 2015; Bonenberger, Aikins, Akweongo, & Wyss, 2014; Delobelle et al., 2011; Heponiemi et al., 2014; Rouleau et al., 2012). The pandemic had drastically affected the economic and social facets of almost all the businesses across all the continents. Several employees had to modify their work pattern and take up newly adopted working patterns such as virtual meetings, flexible working arrangements, and so on. There were several other employees who lost either their job or opportunities for growth and promotion (Nyanga & Zirima, 2020) which resultantly causing them to experience poorer job satisfaction which in turn not only hampered their job performance and work commitment but had also undermined their psychological well-being and caused them to suffer from occupational stress, anxiety, and depression. According to previous evidence, resilience attenuates the debilitating effects of stress on job satisfaction (Krush, Agnihotri, Trainor, & Krishnakumar, 2013). Employees with high resilience have the capability to recover themselves from the negative emotions, flexibly adapt to the changing demand of stressful situations, and effectively cope up with stressful situations in comparison to their less resilient peers. Previous researches have manifested a weak to moderate positive relationship between job

satisfaction and resilience (Hudgins, 2016; Larson & Luthans, 2006; Matos, Neushotz, Griffin, & Fitzpatrick, 2010). The governmental imposed lockdown due to pandemic has radically changed the working pattern and forced employees to WFH. Workers with lower level of resilience found it difficult to accept and adapt the new normal, which culminated poor employee performance, work dissatisfaction, and poor work life balance and altogether this may hamper the mental or emotional well-being as well.

3. **Work Engagement and Resilience:** Next psychological attribute which is undermining due to pandemic enforced digital working is work engagement. Schaufeli et al. in the year 2002, defined employee work engagement as "positive, fulfilling, work-related state of mind that is characterized by vigor, dedication, and absorption." According to Rich et al. (2010) engaged employees are more attentive, more focused on their responsibilities, more enthusiastic in comparison to those of less engaged workers, and other researchers suggested that they are more likely to strengthen to sustainable individual and organizational development while encouraging a healthy workplace (Bakker et al., 2011) because engaged employees enthusiastically participate in social activities, hobbies, and outside work (Schaufeli et al., 2001), positive effects of work engagement affect the private life and vice versa (Grzywacz & Marks, 2000), which in turn leads to enhanced individual and group goal attainment. During turbulent times, like the COVID-19 pandemic, the employees are finding it difficult to focus on their job. The threat of oneself or family members getting infected from the virus, remote working, work-life conflict, job insecurity, long working hours, social isolation are disengaging them from their work. These stress-related factors are generating psychological strain and are deteriorating their mental health and well-being. Researches have shown that employees possessing resilience experience positive emotions and are being able to deal with stress and give room to concentrate on job task (Philippe, Lecours, & Beaulieu-Pelletier, 2009). Employees who manifest positive emotions and resilience, irrespective of adverse situations, are energetic and depict an affective connection to perform their job duties. Bande, Fernández-Ferrín, Varela, & Jaramillo (2015) noted that resilience could lead to work satisfaction and

engagement. Because engaged employees have energetic and affective bond with their job related activities, find themselves as capable of managing with job related demands (Bakker, 2009; Demerouti et al., 2010), they are more likely to strengthen growth and development at individual and organizational front while uplifting a healthy work environment (Bakker, 2011).

4. **Work Family Conflict and Resilience:** Greenhaus & Beutell (1985) defined work-family conflict, a form of inter-role conflict, arises due to incompatible time, energy, and behavioral demands of work and family domains. Work-family conflict is a psychological phenomenon of imbalance between work and home life (Csikszentmihalyi, 2003). Among the various non-pharmaceutical guidelines released by the regional and national governments across the continents, one was to work remotely. The employees engaged in physical mode work initially considered application of the new normal as one of the measure to handle the work family conflicts as the required time to commute was saved, they can work and manage the house hold responsibilities simultaneously. But due to prolong lockdown, employees faced work family conflict due to long working hours, getting the calls from the managers after the work hours as well, the webinars and meetings were organized, which were creating stress among the employees. Not only was this the role conflict, there were issues from the family front also. The children were not going to schools and tuitions, so along with their office work, they have to manage the academic and extra-curricular activities of their children (Rudolph, 2020). Since all these were occurring simultaneously and were incompatible in nature, hence, it hampered the employees' psycho-emotional state and cause tension (Lonska et al., 2021), depression and anxiety. It may also negatively affect the physical health including poorer sleep and perceived health, as well as higher cardiometabolic risk (Perrewe & Ganster, 2006). In addition to it, they may also tend to get engage more in unhealthy behaviors, such as smoking, alcohol use, medication overuse, poor diet, and sedentary behavior (Greenhaus, Allen, & Spector, 2006). Consequently, the companies may experience deterioration in work motivation, work efficiency, production rate, and loyalty of the employees (Johnson et al., 2005). Evidence from researches

have proved that employees having high resilience can manage the work family conflict in a better way than those with lower. Siu in 2013 noted that psychological resources such as optimism and resiliency are related to increased work-life balance. Chen, Li, Xu, & Han (2021) through their research found that nurses' resilience can help them in scale down the level of work-family conflict. Employees with resilience have capacity to cope up with stress generated due to work family conflict amid pandemic. They can handle the negative emotions and bounce back and attain better psychological well-being.

5. **Work Performance and Resilience:** Another organizational psychology attribute which not only determines the success of a company (Shaughnessy, 2017) but is also important for the employees as achieving goals or tasks can be a source of satisfaction (Muchhal, 2014) is work performances. It is defined as the way through which an employee contributes to organizational goals (Campbell, 1990). Atatsi et al. in the year 2019, defined work performance as the extent to which the employees fulfill the demand of task assigned in an effectively managed manner within the given resources. It majorly consists of three domains: (i) task performance; (ii) contextual performance; and (iii) adaptive performance (Koopmans et al., 2011). *Task performance* is related to those activities of an employee which undertake on-duty and is straightforwardly related to the particular job profile. *Contextual performance* is defined as those actions which surpass the officially described work objectives (Koopmans et al., 2011). *Adaptive Performance* is about employee learning new things in their workplace. It reflects an employee's capacity to handle the changes that affect their position as a worker.

The employee's work performance depends on several factors such as equipment, physical work environment, meaningful work, standard operating procedures, reward for good or bad systems, performance expectancy, feedback on performance, knowledge, skills (Stup, 2003) training, security/safety, working hours and job redesigning (Kiazad et al., 2019; Masadeh et al., 2016) and mental health (Sasaki et al., 2020).

Amid turbulent phase, like pandemic, it has been noted that working from home has reduced the performance level of employees. Frustration, work family conflict, digital inequalities, stresses, not all work can be done from home and many issues affected the work performance of employees (Wang et al., 2020). In 2018, Jackson & Frame argued that employee's level of performance improves with their capacity to deal with stressful events up to a certain extent. However, after reaching that point, increased work stress deteriorates the work performance. Precedent research found a positive association between resilience and job performance (Luthans et al., 2007). In the context of COVID-19, it's likely that employees' resilience would interact with their personality to shape their psychological well-being, which in turn predicts their task and relational outcome at work (Charoensap-Kelly, Sheldon, Antony, & Provenzani, 2021).

As stated above, the sudden transition to curb the spread of the virus has not only deteriorated the work efficiency, but has also affected job performance. It has also caused them to suffer from occupational burn out, stress which in turn affects their psychological well-being and causing them to suffer from psychological ailments such as anxiety, stress, and depression. In the absence of healthy emotional and mental well-being, an employee cannot prove to be a beneficial resource for the organization which will impact its growth and economy. With the collaborative efforts of employees and companies, psychological, and mental well-being of employees can be fostered. Following are some the suggestive measures which can be considered to mitigate the psychological impact of remote working among employees amid pandemic:

1. **At Employee's Level:**
   i. By strengthening family bonds and interaction. Through researches, it has been observed that the family bond is inversely related to psychological ailments such as stress, anxiety, and depression. Those family members who spend together and participate in family interactions are found to be having better psychological well-being. Stronger family rituals and routines are predictive of family attachment and cohesion which may act as a core factor of family resilience during difficult times (Harrist, 2019).
   ii. The research on resilience suggests that individuals learn new strategies for managing work-related and private requirements during the COVID-19 pandemic, which can increase their specific self-efficacy beliefs dealing with future crises (Rudolph et al., 2020).

iii. Employees should be acknowledged for their performance; such recognition will boost their confidence in making efforts to attain goal, meet deadlines even during challenging situations.
iv. Employees should learn how to manage their time effectively so that they can play a proactive role at personal and professional front. They should create a fine line between the two.
v. They must develop self-efficacy in themselves as it is going to bring confidence in them to carry out the task assigned virtually. With the increase in level of self-efficacy the work motivation and job performance will also increase.
vi. In order to mitigate the psychological impact of digital working (such as occupational stress, e-work family conflict, job dissatisfaction), employees must practice yoga and meditation. Practicing such activities will be going to help them in deal with turmoil like COVID-19 and have effective psychological well-being.

2. Organizational Level:

i. The organization must consider about mental health of its employees. Working remotely, managing household, taking of children and their online classes and work chores simultaneously generate stress and anxiety. Hence, organization should arrange some online stress management programs or yoga sessions for its employees which can help them in handling negative emotions and attaining a better psychological well-being.
ii. To promote work efficiency of employees while WFH, organizations should be flexible in terms of work performance. This flexibility could help them in juggle between assigned task and home related responsibilities. This will also help them in spending quality time with their near and dear ones, thereby making them experience lesser psychological distress by sharing their worries and experiences with them.
iii. The organization should provide some facilities which can improve the work productivity. The facilities include efficient IT infrastructure for easy access to relevant databases and

information from remote locations. Facilities need to be user friendly and secure to encourage employees' acceptance.

iv. Companies should safeguard employees by providing health and safety care insurance schemes. This will aid in assuring the health and safety of employees and their family members, while at the same time motivating employees to commit and engage towards their work.
v. By recognizing and appreciating the task accomplished by employees while working remotely can also boost their confidence, promote work motivation, and job engagement.
vi. The companies should not impose long working hours especially amid COVID-19 virus outbreak. As it will affect their work motivation and reduced the work performance.
vii. The companies should hire psychologist who through tele-counseling services can empathize with them and can help employees in handling psychosocial issues caused due to pandemic enforced digital working and motivate them to work effectively and efficiently for the company.

6.3 CONCLUSION

COVID-19 has devastatingly affected everyone across the globe. To mitigate its effect, governments of different countries have imposed lockdown in one or the other way which compelled all of us to adapt a 'new normal' life style. This new normal has detrimentally affected the physical, psychological, social, and emotional health of an individual. Employees belonging from different occupational sectors have faced different sort of challenges amid the pandemic. Occupational sectors such as tourism went down on the other side individuals belonging from health care sector were having heavy work load and burn out which was a heavy toll on them and deteriorated their mental and emotional well-being. The existing chapter highlighted whether resilience can help in predicting the psychological well-being of the employees working remotely. Through extensive review literature, it was noted pandemic has caused devastating effect on the psychological well-being of employees. Sitting in front of screens for longer periods of time, attending virtual meetings after working hours, doing assigned tasks, taking care of children and elderly people, and doing household chores were going hand in hand. All these were compromising their psychological health as well as work efficiency. The existing chapter focused on how employees working amid pandemic encountered deterioration on various psychological attributes such as work motivation, work family conflict, job satisfaction. Through precedent researches it was noted that resilience was inversely related to stress generated due to pandemic and remote working. This means that those employees possessing resilience were being able to manage the pandemic stress and maintain emotional and mental health. Resilience can have dual benefits, on one end; it may help employees in accomplishing the assigned tasks, maintain effective work life balance and lead a mentally peaceful life. On the other end, having a manpower resource, that is resilient in nature, can contribute in the growth and development of the organizations. Hence, in order to reduce the impact of pandemic on psychological well-being of employees there is a need to enhance resilience among them which can be done with the collaborative efforts of employees and companies. To build resilience amid pandemic the employees can create a boundary between work and family, practice yoga and meditation, learn to manage time, on the other hand, companies may organize webinars/workshops on stress management, yoga. They may also be flexible in terms of time, recognize the efforts and accomplishments of the employees. In nutshell, it was found that remote working, amid pandemic, has caused detrimental effect on psychological well-being of the employees but positive psychological traits such as resilient can help them to fight with such turbulent time and can prove to be a beneficial manpower resource to the company.

KEYWORDS

- **economic crisis**
- **pandemic**
- **psychological well-being**
- **resilience**
- **severe acute respiratory syndrome corona virus**
- **tele-counseling**
- **work from home**
- **yoga**

REFERENCES

American Psychological Association, (2005). *The Road to Resilience.* [Internet] https://uncw.edu/studentaffairs/committees/pdc/documents/the%20road%20to%20resilience.pdf (accessed on 7 February 2023).

Atatsi, E. A., Stoffers, J., & Kil, A., (2019). Factors affecting employee performance: A systematic literature review. *J. Adv. Manag. Res., 16*, 329–351. doi: 10.1108/JAMR-06-2018-0052.

Avey, J. B., Luthans, F., & Jensen, S. M., (2009). Psychological capital: A positive resource for combating employee stress and turnover. *Human Resource Management, 48*(5), 677–693.

Avey, J. B., Luthans, F., Smith, R. M., & Palmer, N. F., (2010). Impact of positive psychological capital on employee well-being over time. *Journal of Occupational & Health Psychology, 15*(1), 17–28.

Aziri, B., (2011). Job satisfaction: A literature review. *Management Research & Practice, 3*(4), 77–86.

Bakker, A. B., (2009). Building engagement in the workplace. In: Burke, R. J., & Cooper, C. L., (eds.) *The Peak Performing Organization* (pp. 50–72). Milton Park: Routledge. doi: 10.4324/9780203971611.ch3.

Bakker, A. B., Albrecht, S. L., & Leiter, M. P., (2011). Key questions regarding work engagement. *Eur. J. Work Organ. Psychol., 20*, 4–28. doi: 10.1080/1359432X.2010.485352.

Bande, B., Fernández-Ferrín, P., Varela, J. A., & Jaramillo, F., (2015). Emotions and salesperson propensity to leave: The effects of emotional intelligence and resilience. *Industrial Marketing Management, 44*, 142–153.

Bandura, A., (2008). Social cognitive theory. In Donsbach, W., (ed.), *International Encyclopedia of Communication* (pp. 4654–4659). Oxford, UK: Blackwell.

Bonenberger, M., Aikins, M., Akweongo, P., & Wyss, K., (2014). The effects of health worker motivation and job satisfaction on turnover intention in Ghana: A cross-sectional study. *Human Resources for Health, 12*(1), 1–12.

Brooks, S. K., Webster, R. K., Smith, L. E., Woodland, L., Wessely, S., Greenberg, N., & Rubin, G. J. (2020). The psychological impact of quarantine and how to reduce it: Rapid review of the evidence. *The Lancet, 395*(10227), 912–920.

Buchanan, G. M., & Seligman, M. E. B., (1995). *Explanatory Style*. Hillsdale, NJ: Erlbaum.

Business, G., (2003). *Leadership, Flow, and the Making of Meaning.* New York: Viking.

Campbell, J. P., (1990). Modeling the performance prediction problem in industrial and organizational psychology. In: Dunnette, M. D., & Lough, L. M., (eds.), *Handbook of Industrial and Organizational Psychology* (pp. 687–732). American Psychological Association: Worcester, MA, USA.

Charoensap-Kelly, P., Sheldon, P., Antony, M. G., & Provenzani, L., (2021). Resilience, well-being, and organizational outcomes of Croatian, Thai, and US workers during COVID-19. *Journal of Management & Organization*, 1–29.

Chaturvedula, S., & Joseph, C., (2007). Dimensions of psychological well-being and personality in military aircrew: A preliminary study. *Indian Journal of Aerospace Medicine, 51*(2), 17–27.

Chen, X., Li, Q., Xu, F., & Han, B., (2021). The mediating role of resilience between work–family conflict and career development among Chinese nurses: A cross-sectional study. *Journal of Nursing Management, 29*(6), 1733–1741. doi: 10.1111/jonm.13323.

Cho, E., (2020). Examining boundaries to understand the impact of COVID-19 on vocational behaviors. *Journal of Vocational Behavior, 119*, 103437.

Delobelle, P., Rawlinson, J. L., Ntuli, S., Malatsi, I., Decock, R., & Depoorter, A. M., (2011). Job satisfaction and turnover intent of primary healthcare nurses in rural South Africa: A questionnaire survey. *Journal of Advanced Nursing, 67*(2), 371–383.

Demerouti, E., Cropanzano, R., Bakker, A., & Leiter, M., (2010). From thought to action: Employee work engagement and job performance. *Work Engagement: A Handbook of Essential Theory and Research, 65*(1), 147–163.

Duckworth, A. L., Peterson, C., Matthews, M. D., & Kelly, D. R., (2007). Grit: Perseverance and passion for long-term goals. *J. Pers. Soc. Psychol., 92,* 1087–1101.

Duckworth, A. L., Quinn, P. D., & Seligman, M. E., (2009). Positive predictors of teacher effectiveness. *The Journal of Positive Psychology, 4*(6), 540–547.

Evans, L., (1997). Understanding teacher morale and job satisfaction. *Teaching and Teacher Education, 13*(8), 831–884.

Giorgi, G., Lecca, L. I., Alessio, F., Finstad, G. L., Bondanini, G., Lulli, L. G., & Mucci, N., (2020). COVID-19-related mental health effects in the workplace: A narrative review. *International Journal of Environmental Research and Public Health, 17*(21), 785.

Greenhaus, J. H., & Beutell, N. J., (1985). Sources of conflict between work and family roles. *Academy of Management Review, 10*(1), 76–88.

Greenhaus, J. H., Allen, T. D., & Spector, P. E., (2006). Health consequences of work–family conflict: The dark side of the work–family interface. In: *Employee Health, Coping and Methodologies*. Emerald Group Publishing Limited.

Grzywacz, J. G., & Marks, N. F., (2000). Reconceptualizing the work–family interface: An ecological perspective on the correlates of positive and negative spillover between work and family. *J. Occup. Health Psychol. 5,* 111. doi: 10.1037/1076-8998.5.1.111.

Halbesleben, J. R., Neveu, J. P., Paustian-Underdahl, S. C., & Westman, M., (2014). Getting to the "COR" understanding the role of resources in conservation of resources theory. *Journal of Management, 40*(5), 1334–1364.

Harrist, A. W., Henry, C. S., Liu, C., & Morris, A. S., (2019). Family resilience: The power of rituals and routines in family adaptive systems. In Fiese, B. H., Celano, M., Deater-Deckard, K., Jouriles, E. N., & Whisman, M. A., (eds.), *APA Handbook of Contemporary Family Psychology: Foundations, Methods, and Contemporary Issues Across the Lifespan* (pp. 223–239).

Heponiemi, T., Kouvonen, A., Virtanen, M., Vänskä, J., & Elovainio, M., (2014). The prospective effects of workplace violence on physicians' job satisfaction and turnover intentions: The buffering effect of job control. *BMC Health Services Research, 14*(1), 1–8.

Hudgins, T. A., (2016). Resilience, job satisfaction and anticipated turnover in nurse leaders. *Journal of Nursing Management, 24*(1), E62–E69.

Huppert, F. A., (2009). Psychological well-being: Evidence regarding its causes and consequences. *Applied Psychology: Health and Well-Being, 1*(2), 137–164.

Jadoo, S. A. A., Aljunid, S. M., Dastan, I., Tawfeeq, R. S., Mustafa, M. A., Ganasegeran, K., & AlDubai, S. A. R., (2015). Job satisfaction and turnover intention among Iraqi doctors-a descriptive cross-sectional multicenter study. *Human Resources for Health, 13*(1), 1–11.

Johnson, S., Cooper, C., Cartwright, S., Donald, I., Taylor, P., & Millet, C., (2005). The experience of work-related stress across occupations. *J. Manag. Psychol., 20*, 178–187. doi: 10.1108/02683940510579803.

Kelly, D. R., Matthews, M. D., & Bartone, P. T., (2014). Grit and hardiness as predictors of performance among west point cadets. *Military Psychology, 26*(4), 327–342.

Kiazad, K., Kraimer, M. L., & Seibert, S. E., (2019). More than grateful: How employee embeddedness explains the link between psychological contract fulfillment and employee extra-role behavior. *Human Relations, 72*(8), 1315–1340.

Kirkwood, T. B., Bond, J., May, C., McKeith, I., & Teh, M. M., (2014). Foresight mental capital and well-being project: Mental capital through life: Future challenges. *Well-being: A Complete Reference Guide*, 1–90.

Koopmans, L., Bernaards, C. M., Hildebrandt, V. H., Van, B. S., Van, D. B. A. J., & De Vet, H. C., (2014). Improving the individual work performance questionnaire using Rasch analysis. *J. Appl. Meas., 15*, 160–175.

Krush, M. T., Agnihotri, R. A. J., Trainor, K. J., & Krishnakumar, S., (2013). The salesperson's ability to bounce back: Examining the moderating role of resiliency on forms of intra-role job conflict and job attitudes, behaviors and performance. *Marketing Management Journal, 23*(1), 42–56.

Larson, M., & Luthans, F., (2006). Potential added value of psychological capital in predicting work attitudes. *Journal of Leadership & Organizational Studies, 13*, 75–92. doi: 10.1177/ 10717919070130020601.

Lease, S. H., (1998). Annual review, 1993–1997: Work attitudes and outcomes. *Journal of Vocational Behavior, 53*(2), 154–183.

Li-Ping, T. T., Kim, J. K., & Shin-Hsiung, T. D., (2000). Does attitude toward money moderate the relationship between intrinsic job satisfaction and voluntary turnover? *Human Relations, 53*(2), 213–245.

Lonska, J., Mietule, I., Litavniece, L., Arbidane, I., Vanadzins, I., Matisane, L., & Paegle, L., (2021). Work–life balance of the employed population during the emergency situation of COVID-19 in Latvia. *Frontiers in Psychology, 12*. doi: 10.3389/fpsyg.2021.682459

Lu, Y., Hu, X. M., Huang, X. L., Zhuang, X. D., Guo, P., Feng, L. F., & Hao, Y. T., (2016). Job satisfaction and associated factors among healthcare staff: A cross-sectional study in Guangdong province, China. *BMJ Open, 6*(7), e011388.

Martin, A. J., (2005). The role of positive psychology in enhancing satisfaction, motivation, and productivity in the workplace. *Journal of Organizational Behavior Management, 24*(1, 2), 113–133.

Masadeh, R., Obeidat, B. Y., & Tarhini, A., (2016). A Jordanian empirical study of the associations among transformational leadership, transactional leadership, knowledge sharing, job performance, and firm performance: A structural equation modelling approach. *Journal of Management Development, 35*(5), 681–705.

Matos, P. S., Neushotz, L. A., Griffin, M. T. Q., & Fitzpatrick, J. J., (2010). An exploratory study of resilience and job satisfaction among psychiatric nurses working in inpatient units. *International Journal of Mental Health Nursing, 19*(5), 307–312.

Millar, J. D., (1984). The NIOSH-suggested list of the ten leading work-related diseases and injuries. *Journal of Occupational and Environmental Medicine, 26*(5), 340–341.

Moe, K. (2012). "Factors Influencing Women's Psychological Well-being within a Positive Functioning Framework." *Theses and Dissertations--Educational, School, and Counseling Psychology*, 3.

Muchhal, D. S., (2014). HR practices and job performance. *IOSR Journal of Humanities and Social Science (IOSR-JHSS), 19*(4), 55–61.

Neria, Y., Nandi, A., & Galea, S., (2008). Post-traumatic stress disorder following disasters: A systematic review. *Psychological Medicine, 38*(4), 467–480.

Nyanga, T., & Zirima, H., (2020). Reactions of small to medium enterprises in Masvingo, Zimbabwe to COVID-19: Implications on productivity. *Business Excellence and Management, 10*(1), 22–32.

O'Driscoll, M., & Brough, P., (2003). Job stress and burnout. *Organizational Psychology in Australia and New Zealand*, 188–211.

Ong, A. D., Bergeman, C. S., Bisconti, T. L., & Wallace, K. A., (2006). Psychological resilience, positive emotions, and successful adaptation to stress in later life. *Journal of Personality and Social Psychology, 91*(4), 730.

Ozcelik, H., & Barsade, S. G., (2018). No employee an island: Workplace loneliness and job performance. *Academy of Management Journal, 61*(6), 2343–2366.

Padhy, M., Chelli, K., & Padiri, R. A., (2015). Optimism and psychological well-being of police officers with different work experiences. *Sage Open, 5*(2), 2158244015580852.

Pakenham, K. I., Landi, G., Boccolini, G., Furlani, A., Grandi, S., & Tossani, E., (2020). The moderating roles of psychological flexibility and inflexibility on the mental health impacts of COVID-19 pandemic and lockdown in Italy. *Journal of Contextual Behavioral Science, 17*, 109–118.

Perrewe, P. L., & Ganster, D. C., (2006). *Research in Occupational Stress and Well-Being*, 61–98.

Peterson, S. J., & Byron, K., (2008). Exploring the role of hope in job performance: Results from four studies. *Journal of Organizational Behavior: The International Journal of Industrial, Occupational and Organizational Psychology and Behavior, 29*(6), 785–803.

Philippe, F. L., Lecours, S., & Beaulieu-Pelletier, G., (2009). Resilience and positive emotions: Examining the role of emotional memories. *Journal of Personality, 77*(1), 139–176.

Pillay, R., (2009). Work satisfaction of professional nurses in South Africa: A comparative analysis of the public and private sectors. *Human Resources for Health, 7*(1), 1–10.

Rich, B. L., Lepine, J. A., & Crawford, E. R., (2010). Job engagement: Antecedents and effects on job performance. *Academy of Management Journal, 53*(3), 617–635.

Robbins, S. P., (2005). *Organizational Behavior* (11th edn.). New Jersey: Pearson Education.

Rouleau, D., Fournier, P., Philibert, A., Mbengue, B., & Dumont, A., (2012). The effects of midwives' job satisfaction on burnout, intention to quit and turnover: A longitudinal study in Senegal. *Human Resources for Health, 10*(1), 1–14.

Rudolph, C. W., Allan, B., Clark, M., Hertel, G., Hirschi, A., Kunze, F., & Zacher, H., (2021). Pandemics: Implications for research and practice in industrial and organizational psychology. *Industrial and Organizational Psychology, 14*(1, 2), 1–35.

Sasaki, N., Kuroda, R., Tsuno, K., & Kawakami, N., (2020). Workplace responses to COVID-19 associated with mental health and work performance of employees in Japan. *Journal of Occupational Health, 62*(1), e12134.

Schaufeli, W. B., Martínez, I. M., Pinto, A. M., Salanova, M., & Bakker, A. B., (2002). Burnout y el compromiso de los estudiantes universitarios: Un estudio nacional. *Diario de la Cruz-Psicología Cultural, 33*, 464–481.

Schaufeli, W., Taris, T., Le Blanc, P., Peeters, M., Bakker, A., & De Jonge, J., (2001). Makes Work Healthy. *Looking for the Enthusiastic Employee [Does Work Make Happy].* 422–428.

Schoeman, J. D., (2012). *The Relationship Between Work Place Well-Being, Psychological Capital and Work Place Trust.* Doctoral dissertation, Nelson Mandela Metropolitan University.

Seligman, M. E. P., (1998). *Learned Optimism.* Simon and Schuster, New York, NY.

Shagvaliyeva, S., & Yazdanifard, R., (2014). Impact of flexible working hours on work-life balance. *American Journal of Industrial and Business Management, 4*(1) doi: 10.4236/ajibm.2014.41004.

Shaughnessy, A. O., (2017). *Employees are a Company's Greatest Asset, They're Your Competitive Advantage.* Poppulo. https://www.poppulo.com/blog/employees-are-a-companys-greatest-asset-theyre-your-competitive-advantage (accessed on 7 February 2023).

Shimazu, A., Nakata, A., Nagata, T., Arakawa, Y., Kuroda, S., Inamizu, N., & Yamamoto, I., (2020). Psychosocial impact of COVID-19 for general workers. *Journal of Occupational Health, 62*(1), e12132.

Siu, O. L., (2013). Psychological capital, work well-being, and work-life balance among Chinese employees. *Journal of Personnel Psychology, 12*(4).

Spector, P. E., (1997). *Job Satisfaction: Application, Assessment, Causes, and Consequences* (Vol. 3). Sage.

Stup, R., (2003). *Control the Factors That Influence Employee Success.* Managing the Hispanic workforce conference. Cornell University and Pennsylvania State University.

Temesgen, K., Aycheh, M. W., & Leshargie, C. T., (2018). Job satisfaction and associated factors among health professionals working at Western Amhara Region, Ethiopia. *Health and Quality of Life Outcomes, 16*(1), 1–7.

Usman, A., (2017). The effect of psychological well-being on employee job performance: Comparison between the employees of projectized and non-projectized organizations. *Journal of Entrepreneurship & Organization Management, 6*(01), 8–12.

Wang, C., Cheng, Z., Yue, X. G., & McAleer, M., (2020). Risk management of COVID-19 by universities in China. *Journal of Risk and Financial Management, 13*, 36–41.

Westerhof, G. J., & Keyes, C. L., (2010). Mental illness and mental health: The two continua model across the lifespan. *Journal of Adult Development, 17*(2), 110–119.

World Health Organization, (2003). *Investing in Mental Health.* Geneva.

World Health Organization. *Corona Virus Disease (COVID-19).* Available Online: https://www.who.int/health-topics/coronavirus#tab=tab_2 (accessed on 7 February 2023).

Wright, T. A., (2010). The role of employee well-being in organizational research. *Oxford Handbook of Positive Psychology and Work*, 143–154.

Yavas, U., Babakus, E., & Karatepe, O. M., (2013). Does hope moderate the impact of job burnout on frontline bank employees' in-role and extra-role performances? *International Journal of Bank Marketing, 31*(1), 56–70.

Zhu, N., Zhang, D., Wang, W., Li, X., Yang, B., Song, J., et al., (2020). A novel coronavirus from patients with pneumonia in China. *The New Zealand Journal of Medicine, 382*(8), 727–733.

CHAPTER 7

Impact of the Pandemic on Street Food Vendors Globally: Challenges and Coping Strategies

VISMITA PALIWAL,[1] RUCHI JOSHI,[2] and NIDA NAFEES[3]

[1]*Associate Professor and Coordinator, Amity Institute of Behavioral and Allied Sciences, Amity University, Rajasthan, India*

[2]*Assistant Professor, Department of Psychology, Manipal University Jaipur, India*

[3]*Assistant Professor, Amity Institute of Behavioral and Allied Sciences, Amity University, Rajasthan, India*

ABSTRACT

Street food has always been an integral part of every culture. From mouthwatering spicy and tangy food items to soothing sweet tooth cravings, street food comprises an indispensable part of all cultures across the world. The street food economy boosts countries' tourism and is especially very beneficial for countries that heavily rely upon tourism. Usually, street food vendors comprise of a large chunk of unorganized sector and provide employment to a variety of groups who cannot find otherwise find employment easily, especially rural migrants.

The pandemic and its repercussions have affected every aspect of human life; be it Physical and Mental Health, Academics or Economy and Food industry was no different, specially the Street Food Vendors were hit adversely as well.

Reshaping the Business World Post-COVID-19: Management Strategies for Sustainable Behavior Change. Arvind K. Birdie & Ruchi Joshi, (Eds.)

The present qualitative study attempts to evaluate the Socio-Economic as well as psychological impact on the lives and business of vendors during and post-lockdown, the challenges faced as well as the coping strategies employed to overcome the challenges worldwide through semi-structured interviews conducted on 3 Street Food Vendors from India, Italy and Indonesia. The interviews were recorded and transcript. Three Major themes surfaced from the interviews, namely; Impact on Business and Operations, Psychological Impact and Coping Strategies. The chapter also throws light on some of the implications, if employed can prove fruitful in improving lives of Street Food Vendors who are at the lowest steps of the social and economic ladder, worldwide due to dearth of adequate employment opportunities, educational qualifications, skill set and lack of sympathetic attitude from government as well as civic authorities.

7.1 INTRODUCTION

Susanne Hooper, an Architect from Canada explicitly remembers her chance encounter with local food cuisine on her first ever official trip to Singapore where an impromptu presentation made her leave the Hotel without having her breakfast and left her famished well past Lunch hour. Starving she approached the street food vendor stall and ordered what seemed very vibrant and the aroma promising what she called "food-gasm." Hainanese Chicken Rice, also known as the unofficial national food of Singapore, was pure bliss. *"Not only it was delightfully tasty and filling but also very cheap compared to expensive, bland food served in the high-end Hotel I was putting up."* She ended up eating there, trying new delicacies all through her stay. *"What thrilled me was not just taste bud satisfying food but also an opportunity to learn about the authentic Singapore food and the joy of actually giving business to a local instead of the International Hotel Chain. Food is an integral part of the culture and ethnicity of a place, especially the street food served with so much enthusiasm. What I ate there was not just food but raw emotion, and all of this in less than 20 dollars per meal!"*

After the deadly waves of COVID-19, when travel restrictions were lifted and an extremely cautious Hooper traveled to Singapore for Business eating in her hotel that boosted of double vaccinated staff and was sanitized almost every nano-second, missed not only the juicy prawns, spicy chicken wings, desserts of fried carrot cake and delightful dallops of Ice Kachang spoiling her formal attire, the last time as she fondly remembers but also the joy of trying authentic local cuisine and interacting with the owner and other customers, observing them, sharing giggles and exchanging information about culture and climate of their native places. "*What these interactions over the food taught me, no guided tourism could ever teach,*" she reflects ruefully.

Street food has always been an integral part of indigenous culture of countries and continents in almost all sections of society owing to its easy availability, the satisfaction it provides to taste buds as well as for being economical and easy on pocket. There are numerous Street Food Joints and Venders that are famous for their delicacies throughout the world.

The Food and Agriculture Organization of the United Nations (FAO, 1989) defines street foods as "ready-to-eat foods and beverages prepared and/or sold by vendors and hawkers especially in streets" Often these vendors do not have permanent residential buildings for running business, are not much educated, lack specific employability skills, do not have legal support from administration and frequently harassed by civil authorities.

Street Food Vendors are not only source of nutrition or "eating out on a weekend" but provides employment to a large section of the unorganized sector. Street Food vending plays an extremely important role in providing impetus to economies around the world, especially underdeveloped countries.

The year 2020 took the world by storm when COVID-19 or what is known as the Corona Virus first surfaced in China and soon engulfed all countries and continents across the world. On 11 March 2020, the World Health Organization declared Corona Virus as a Pandemic (WHO, 2020). The world was not prepared for a medical emergency that was hitherto unheard of with shortage of medicines and equipment and overworked medical fraternity, and Governments across the world implemented Lockdowns to check the spread of infection.

While Lockdowns proved to be an effective measure to curb the infection, they proved detrimental to economies across the world. Almost every sector felt the tremors but the worst among them were the ones who were at the lowest step of the ladder, i.e., the workers from the unorganized sector that consisted of blue-collar job holders such as laborers, miners, domestic helps and Street Vendors with Street Food Vending being no exception.

The impact of pandemic was visible on the earnings of Street Food Vendors even before the Lockdowns came into full force as the fear of infection prevented a large chuck of customers to eat at the local joints.

A few places where regular customers, especially those from lower socio-economic background visited regularly were soon evicted, citing the Vending services as a source of spreading infection.

This scenario played out in different parts of the country and world where the Street Food Vendors bore the brunt of economic loss and social stigma simultaneously.

The present chapter attempts to investigate the impact of pandemic on Street Food Vending globally and strategies employed by governments at the collective level and by the vendors personally to overcome the predicaments they were caught unaware of.

In a study conducted by Romero-Michel et al. (2021) in Mexico, it was proven that street vendors were a sector of the population that was highly vulnerable to significant economic loss due to contingency measures. Informal workers couldn't stop working despite the "Stay at Home" initiative because the government did not implement strategies that would guarantee their survival and their families. Therefore, street vendors continue to be a source of the virus's spread throughout cities.

Kiaka et al. (2021) in their collaborative and comparative study on Food Street Vendors of Harare, Zimbabwe, Kisumu cities of Kenya through interviews argued that trading for Street Food Vendors before the pandemic was not easy because of living in contested spaces or what was known as: "Street Geographies" resulting in violence and repressions due to politics of municipalities of the area as well as policies of government. The troubles, however, intensified post-pandemic as street vendors were harassed, streets were swept clean and hygiene standards were imposed. COVID-19 posed, redefined, and created new "street geographies," leaving Street Food Vendors more uncertain and vulnerable than ever.

In another study conducted by Oloko & Ekpo (2021), the effects of COVID-19 lockdown on livelihood and food security on Street Food Vendors and consumers in Nigeria were assessed· in which vendors reported losing income, customers, customer trust and social connection, while consumers reported difficulties in meeting their food and groceries requirements.

Dhillon & Kaur (2021), in their intra-industry analysis of 80 food outlets, investigate whether the impact of COVID-19 on employment level, clientele rate, liquidity constraints, and sustainability aspect of different food outlets is symmetric or asymmetric in nature in Amritsar Punjab found out that bigger food outlets had more potential to survive for longer if lockdowns are imposed again compared to Street Food Vendors owing to their financial availability and stability.

Geetha (2020), in her study on impact of pandemic on Street Food Vendors in Bengaluru, concluded that Street Food Vending provided opportunities for employment in informal sector due to lack of work in organized sector owing to low skills sets, illiteracy, and inability to invest in capital, however, the pandemic pushed many Street Food Vendors to brink of insolvency and going to back to underdeveloped places they had migrated from.

The Researchers interviewed three vendors from India, Indonesia, and Italy to explore the trials faced by them as well as the coping strategies for the same. Informed consent was obtained by all three of them and they were debriefed about the purpose and implications of the study.

The interview shed light on the trials experienced by the owners of food vendors and the approaches adopted by them to overcome the situation.

Street Food has always been an integral part of economy as well as culture of Italy. An article from La Repubblica (2019) stated that the street food sector was worth 19 million in Italy, out of which more than 3 million is in Campania and Emilia Romagna, and over 2 million in Lazio, Sicily, and Lombardy. A conducted by Coldiretti (2013), there were around 2.5 billion street food consumers in the world, 35 million Italians of which 73% consume street food out of habit and 45% choose local specialties, 24% international delicacies whereas about 4% selected traditional products (Privitera & Nesci, 2015). Italian Street food is also one of the major attractions of tourists visiting the country. Italy was one of the worst hit European countries to be affected by pandemic, the effect on Food industry was catastrophic, yet it was the street food vendors who were hit the hardest due to lack of capital and infrastructure.

Tourism contributed 3.2% in economy of Indonesia, compared to 5.9% in 2019. The down fall was brought about by the pandemic. Not only was overall economy affected due to decline in tourism, so was the food industry specifically the Street Food Vendors who are mainly dependent on tourists; both local as well as international.

Gaffar et al. (2022) through their qualitative study in Indonesia conducted by interviewing Street Food Vendors through semi-structured interviews concluded that street food vendors experienced a steep decline in sales and profit, difficulties in procuring raw materials, sustaining business, paying salaries to supporting staff and shifting to new business that could ensure profits.

7.2 METHODOLOGY

The present study is qualitative and descriptive in nature and was conducted through in-depth interviews with the help of a semi-structured interview schedule developed by the Researchers, held with 3 Street Food Vendors that lasted about 30–40 minutes each. The Researchers were from India and the Indian vendor interviewed face to face while other vendors were contacted via video calls and took place in phases.

The interviews were conducted after obtaining verbal informed consent of the vendors, and they were assured strict confidentiality. Also, the interviews were conducted keeping into account that the vendors were not disturbed during active business hours and at their convenience and were free to not answer any question were reluctant to answer or to stop the interview at any point of time if they wanted to.

The following points were taken into consideration regarding the demographic details of the participants who consented to be a part of the study:

- Type of food products sold;
- Age;
- Religion;
- Educational qualification;
- Marital status;
- Number of family members/dependents/children;
- Number of working family members;
- Customers.

7.3 SOURCES OF DATA

Participants were asked questions from demographic questionnaire prepared by researchers, and responses were duly filled. As Semi-structured interview schedule was administered to participants, based on research questions to attain a deeper understanding of participants experiences during the pandemic.

7.4 RESEARCH QUESTIONS

- Who are the customers? Locals or tourists?
- What were the experiences of vendors during, and post-pandemic related to repercussions on business and operations?
- What were the psychological effects on street food vendors during and post-pandemic?
- What were the preventive measures adopted by vendors during and post-pandemic to avoid infection and to ensure safety to self and customers?
- What long-term-preventive measures, according to vendors, must be adopted in case of such medical or other emergencies?

7.5 PROCEDURE

The snowballing technique of sampling was employed to conduct the interviews. The participants were consented and debriefed about the procedure. The interviews were recorded, and the responses were audio-taped and transcribed later. The transcripts were analyzed by identifying relevant, repetitive, non-overlapping statements about how the participants experienced the topic and themes were extracted on the basis of the same.

7.6 RESULTS

After the transcripts, three main themes came, namely, business and operations, psychological impact of pandemic and coping strategies. The descriptions of the three themes is as in subsections.

7.6.1 THEME 1 – BUSINESS AND OPERATIONS

Nitin*, a vendor from India, sold boiled chickpeas and bread made of all-purpose flour in New Delhi. His family owns a piece of land in his native village. The piece of land is been divided between generations and does not suffice to bear expenses for the family with a large number of men working on field and resulting in hidden unemployment. He migrated from Muzaffarpur, Bihar 10 years back due to a lack of employment opportunities and decided to support his family from afar. Before the pandemic hit, he earned 20,000–25,000 Indian Rupee (262-327 United States Dollars). He mostly caters to local customers.

He is married and has four children, parents along with his brother and his family of eight. He had been working in west Delhi at a fixed spot. During and post-pandemic, he reported a steep decline in the sale of food items. He did not have the facility to store left over food items to avoid contamination and getting stale. The items that were not sold, therefore had to be discarded, sold at either no profit no loss state or at losses altogether.

> *"Usually, I would stand at my spot at around 5:30 in evening and ended up selling most of the product. However, just before and during pandemic, customers stopped coming and I was left with my complete ware untouched. During scorching summers boiled potatoes and chopped onions would get spoiled and I was left with no option but to throw them away. I had to take raw materials on credit in the hope of sales, but I was almost bankrupt by the end of second wave. Even post-pandemic footfall of customers is very limited and with extremely heavy heart I had to let go off my assistant of five years."*

Reduced number of hours of operations was another sub-theme that was identified among most of the vendor. Due to curfews imposed where markets were permitted to be operated during specific hours only, the hours for operation of business were reduced drastically that led to a decline in sales, consequently profits.

> *"As it is the pandemic broke our backs. Even when the lockdown was lifted, the weekend curfews made it very difficult to operate. Police vehicles would be on rounds and we had to close the operation in such a hurry like we were some criminals involved in illegal activities."*

> *"I wish someone would teach me how to sell my stuff online. But it looks like companies are only considering the big shots with fancy buildings and furniture inside. Here I sanitize and wash utensils with hot water to ensure safety right in front of everyone's eyes, but customers would still buy from the restaurants only, not us who have been selling healthy, affordable food cooked hygienically. The authorities also chase us away. Ask them will the daily wage earners, construction site workers, masons go to eat in the fancy eateries. Will they order from phone? They will be starved without us, and we will be finished without customers buying from us." Laments Nitin, getting emotional.*

The business and operations were adversely affected, thereby leaving the vendors hopeless and extremely worried for their future and that of their family members. The pandemic surely resulted in customers avoiding eating street food and opting instead for online deliveries which they considered safer.

Abdul Rehman Khan * is a Dim sum seller in Bogor, Indonesia. He has been selling food items from 2015, after he dropped out of academics to support his family. He is married, has 3 children and parents to look after. Prior to pandemic he would make at least 4,705,927 Rupiah (327 Dollars) a month. This amount also included payment to his assistant he had hired, who agrees to work on lower wages. His major business came from customers who were both local as well as international.

> *"The Pandemic led to loss of business. The tourists both indigenous and international stopped coming. The Lock down struck us like lightening. For a long period of time there was no business, the meager savings were depleted. I was unable to pay the little money I was paying to my assistant, and he had to back to his village."*
>
> *—Rues Khan*

> *I am not techno-savvy, I am a simple man. I don't understand how orders are taken online, how payments are made online. My 14-year-old is helping me managing finances, but it is also making him divert his attention from his studies.*

Reduced hours of business was also mentioned by Khan. *Slowly when unlocking started, we were still hopeful. Expecting earlier revenues but authorities did not let us run the business for ample time. Action was taken against us, our ware thrown if continued standing for a few extra minutes.*

Diego* sells Melanzane Parmigiana, a dish that is baked aubergine with cheese along with eggplants. It is one of the traditional cuisine of Italy. Diego who is 24 years old, had started his own business of Street Food Vending in 2019, after having assisted a seasoned chef in one of the successful cafes in Rome. His stall is stationed in Venice, and his customers are mostly foreigners touring Italy. He is separated and could not receive education beyond the elementary level. He only speaks Italian, and his interview had to be translated in English with the help of a translator who comprehends, speaks, and translates both English and Italian. His mother assists him most of the time.

"It came as a sudden blow, there were rumors of a deadly disease that spread from human-to-human transmission in China, we thought it was China's problem. There were tourists in the city. They would buy from us mostly instead of going to the expensive cafes. We sold it for less than 10 Euros (10.98 USD). The evenings were very busy for me and my mother. But when it (virus) started spreading, we practically had to close down the business. The raw material went to waste. We were all asked to close immediately. We live in a rented space. All my savings were spent in paying for rent and food." informs Diego.

"When the lockdown was lifted there limited hours for business but no buyers. My main business came from tourists. The locals could bake it, and most people preferred eating at home after pandemic. Even when the world is crawling back to normalcy, people haven't let down their guards. I tried looking for a job at my previous employer but there are no jobs. I may have to change the business soon. I don't expect too much support from the government."

All three vendors experienced a sharp decline in income, decreased business hours, difficulty in procuring and storing of raw materials, inability to pay salaries to staff and manage daily expenses.

7.6.2 THEME 2 – PSYCHOLOGICAL IMPACT

Not only were the vendors worried about sales, business, and operations they were also worried about their own health. The newspapers, news channels, rumors went into frenzy reporting goriest of details related to spread of infection and fate of those who were infected. The reactions of vendors were spread into two extremes. Majority of them turned into fatalists, exhibiting what Rotter () identified as external locus of control and exhibited careless behavior related to COVID appropriate behavior. Others came out with appropriate precautionary measures but were scared about catching an infection.

"I would go outsell my stuff everyday hearing stories of cremation grounds lined up with relatives of people who died. The man who sold me vegetables passed away as he was infected. Once he was infected, his entire family was also infected. So many people I knew, shared good times with went away. Every day I would think if I got infected, will my family members be able to come? Will I get a decent departure?" He wonders.

The fear of getting infected, of not recovering, of not being able to meet family members and not getting appropriate last rites along with the atmosphere of uncertainly and fear was anxiety provoking. The news of medicines and protective equipment falling short, doctors being over worked, catching infection and succumbing to infection resulted in vendors experiencing stress and anxiety that resulted in mental health issues getting aggravated. The prime concern of government and administrative machinery was on physical health and well-being of people with mental health getting the least priority resulted in mental health of the society in general and vendors specifically getting impacted in a negative manner.

> *"I was very scared for my family my wife who is expecting our fourth child, my young children, my elderly parents. Every time I came home, I was praying to almighty I don't bring home the killer. There were bodies waiting to be buried. It was extremely scary; everything was so uncertain." Khan explains.*

We both were infected, me and my mother. It was my mother I was most worried about. There was a shortage of beds, equipment, doctors, and paramedics. I was isolated at home but my mother's Oxygen levels dipped and she had to be hospitalized for 5 days. Those were scariest 5 days of my life. But we managed, we were lucky that we survived. But this… (he gestures to deserted street of Venice his cart is positioned) *This isn't that lucky.*

The worries about what would happen to one's family if they caught infection had impacted the Vendors the most, Psychologically, some of the symptoms they reported matched a lot with acute and prolonged anxiety. The cognitive abilities declined. They were unable to make decisions, were confused and couldn't think logically.

Many of them experienced distress due to uncertainty of how long the situation would persist. Not just their livelihood but their lives were at stake as well. It can be concluded that pandemic affected the Vendors Psychologically in an adverse manner.

7.6.3 THEME 3 – COPING STRATEGIES

The vendors however did not just sit ideal as mere spectators. As soon as the lockdown was lifted and curfew timings were eased, most of them unable to find alternative employment came back to former business with certain measures to win back the trust of their customers.

> *"As soon as the lockdown was lifted, I ensured that my bottle of disinfectant was visible to all, I have strictly banned everyone from sitting idle and creating unnecessary gatherings. Not everyone listen though, but one must try"*
>
> *—quips Nitin*
>
> *"I have bought soaps and sanitizers for my customers as well I request them put on their masks, to avoid unnecessary gatherings. Sometimes some customers listen but mostly they overlook caution. I cannot force them or be rude to them as this attitude will chase them away."*
>
> *He maintains:*
>
> *"After the Lockdown was lifted, I have also started home deliveries and my wife and parents are selling masks. "I ensure I take a bath as soon as I enter the house, sanitize the open surfaces and encourage the customers to follow COVID appropriate norms." We are also looking forward to backing from Government, I wish civic authorities would extend their support to us as well, instead of chasing us away ruthlessly." Informs Khan.*
>
> *"I would pray. Pray to almighty for mercy. I started attending Masses regularly after we could go to Church." Maintains Diego. "I have started looking for alternative employment. I don't think the cart can manage my expenses any more. I may go back to Rome to look for work."*

The coping strategies included sanitization process and activities in full public view, encouraging customers to follow social distancing norms, looking for other employment opportunities, substance abuse and turning to spirituality were some of the coping strategies employed by Vendors.

7.7 KEY FINDINGS

- The stern actions being taken by the vendors is still unable to inspire complete confidence from the customers.
- The sales and haven't picked up pace post-pandemic, and the new variants tend to affect the business raising uncertainty.

- The demand for their products is been affected adversely because most of the people are still preferring to work from home (WFH) and unemployment has risen significantly.
- Most of the customers are preferring to order online food or are willing to dine out in restaurants and hotels with seating arrangements with the possibility of maintaining social distancing.
- The pandemic had adversely affected the physical, psychological, mental, and financial well-being of vendors.

7.8 CONCLUSION

The present study was undertaken to assess the impact of pandemic globally on street food vendors from three street food vendors from India, Indonesia, and Italy. The study was descriptive and qualitative in nature. Semi-structured interview schedule developed by researchers were administered to the participants after obtaining verbal informed consent. All participants were males, out of which 2 were married and one was separated. Those who were married had 5–9 family members. Most of them also lived in different cities and states and were not from the place from where they operated business. The major themes that came to fore were difficulty in business and operations, adverse psychological impact of pandemic on mental health of vendors and negative coping strategies that included procuring credit on high rates and substance abuse. They had perceived experience of apathy from government and administrative machinery. The key findings reflected that during and post-pandemic period have been very difficult for the vendors as they experienced low profits, short operating hours, low footfall from customers as they prefer eating at home ordering from eateries and restaurants with established buildings, capital, and infrastructure. Most of them believed that only government and administration could help them, but only if they were seriously willing to do so. Majority of them expected government to help them financially by providing loans on low rates of interest and a sympathetic attitude and also priority in vaccination.

The results suggest that condition of Street Food Vendors is almost the same all over the world. These vendors were worst hit because of non-availability of capital, both in terms of infrastructure as well as finances. While governments were busy checking the spread of infection, the people whose livelihood depends on social contact were poorly hit. It will take plenty of time for recovery; for Economies in general and Street Food Vendors in general.

7.8.1 IMPLICATIONS

1. The present study will be helpful in understanding the effect of a pandemic on street food vendors who cater to the nutritional and food requirements of a large section of society, especially those from lower socio-economic backgrounds.
2. Street food vendors constitute a large section of unorganized labor, hence if the vendors are hit badly, the repercussions will be evident in adverse impact on the economy as well.
3. Suggestions by street food vendors for management for business and operations during pandemic will be effective in the long term.
4. The study will be helpful in providing effective mental health care for street food vendors, some of whom admitted having adopted unhealthy coping strategies like substance abuse after experiencing stress, anxiety, and restlessness during present times.

KEYWORDS

- **COVID-19**
- **Food and Agriculture Organization**
- **methodology**
- **pandemic**
- **street food vendors**
- **stress**
- **World Health Organization**

REFERENCES

Coldiretti, (2013). *Coldiretti* Crisi: Boom Cibi di Strada Per 3 Italiani su 4, da Arancini a Piadina. https://www.coldiretti.it/archivio/crisi-boom-cibi-di-strada-per-3-italiani-su-4-da-arancini-a-piadina-2 (accessed on 7 February 2023).

Dhillon, D. K., & Kaur, K., (2021). Impact of COVID-19 on food outlets: Symmetric or asymmetric? A case study of Amritsar. *Journal of Hospitality and Tourism Insights*. https://doi.org/10.1108/JHTI-07-2021-0180.

FAO, (1989). Street Foods. Report of an FAO Expert Consultation, Jogjakarta, Indonesia (Vol. 46, pp. 1–96). FAO Food and Nutrition Paper.

Gaffar, V., Tjahjono, B., Abdullah, T., Sari, M., & Rofaida, R., (2022). Unfolding the impacts of a prolonged COVID-19 pandemic on the sustainability of culinary tourism: Some insights from micro and small street food vendors. *Sustainability, 14*(1), 497. MDPI AG. http://dx.doi.org/10.3390/su14010497.

Geetha, J. R., (2021). Impact of COVID-19 on the street food vendors.: An empirical study. *International Journal of Advanced Academic Studies, 3*(1), 94–99.

Kiaka, R., Chikulo, S., Slootheer, S., et al., (2021). "The street is ours." A comparative analysis of street trading, COVID-19 and new street geographies in Harare, Zimbabwe and Kisumu, Kenya. *Food Sec., 13*, 1263–128. https://doi.org/10.1007/s12571-021-01162-y.

La Repubblica, (2019). *La Repubblica* Il Cibo Conquista le Strade Italiane: Lo Street Food Vale 19 Milioni. https://www.repubblica.it/economia/rapporti/osserva-italia/mercati/2019/08/08/news/il_cibo_conquista_le_strade_italiane_lo_street_food_vale_19_milioni-232592373 (2019) (accessed on 7 February 2023).

OCED, (2020). Indonesia, Tourism in Indonesia. https://www.oecd-ilibrary.org/sites/6c4fc1cd-en/index.html?itemId=/content/component/6c4fc1cd-en#:~:text=Tourism%20has%20boomed%20in%20Indonesia,representing%2010.5%25%20of%20total%20employment (accessed on 7 February 2023).

Oloko, M., & Ekpo, R., (2021). Effects of the COVID-19 lockdown on the livelihood and food security of street food vendors and consumers in Nigeria. *Journal of African Studies and Development, 13*(4), 106–114.

Privitera, & Nesci, (2015). Globalization vs. local. The role of street food in the urban food system. *Procedia Econ. Finance, 22*, 716–722.

Romero-Michel, et al., (2021). Health and economic measures in response to the COVID-19 pandemic – effect on street vendors. *The Journal of Infection in Developing Countries, 15*(2), 198–203. doi: 10.3855/jidc.13465.

World Health Organization, (2020). *WHO Director-General's Opening Remarks at the MEDIA Briefing on COVID-19.*"

CHAPTER 8

Work and Life Allies or Enemies: A COVID Pandemic Perspective

VARSHA SHARMA

Assistant Professor, Department of Psychology, Mohanlal Sukhadia University, Udaipur, Rajasthan

ABSTRACT

'Work and Life' are two important aspects of everyday life which sometimes go together with equal sharing, sometimes overlap, and sometimes interfere or create hindrance in each other. The notion is whether they ally or enemies! This debate is not new, but the COVID pandemic era has added new dimensions in this debate. The term 'work from home' is the most popular one these days. After the emergence of global health emergencies due to COVID, a dramatic change is being experienced in the traditional way of work life balance. This 'new normal' era has come up with lots of opportunities and challenges as well! Now when we are learning to live with COVID and at the same time we are also supposed to learn new work-life balance strategies in order to gain the goal of overall well-being including mental peace and quality of life. Major challenges perceived during COVID were loss of job, loss of near and dear ones, financial insecurity, anxiety for future, health concerns, stressful environment, excessive work hours, low-income jobs and many more. The situations became more problematic during lockdown due to the restricted movement, lack of personal space, non-cooperation from family members, faulty communication patterns, unhealthy family relations, perceived biased behavior of colleagues, unrealistic expectations, etc. Now the need is to identify the

Reshaping the Business World Post-COVID-19: Management Strategies for Sustainable Behavior Change. Arvind K. Birdie & Ruchi Joshi, (Eds.)

ineffective strategies and replace them with the effective strategies. The effective strategies include healthy relationships, tonic habits, assertive communication styles, mindfulness, mental health, practicing strategies for enhancing psychological well-being, etc.

8.1 INTRODUCTION

Work and life balance which is also known as 'Co-Life, has an important role in our lives globally. It is general as well as particular phenomena. The term 'work' indicates the activities which are being performed in order to gain the optimal level of the job performance whereas the other connected term 'Life' indicates the subjective areas of satisfaction in personal life indicating the overall well-being which broadly includes the relationship building and maintenance as well. Work and life balance are two important aspects of life which can collaborate, compete or create chaos. They can bring harmony or disharmony in one's life. Broadly work segment of life includes the business, job targets, financial matters, career, money making, level of job performance, job satisfaction, maintaining the professional relationships, etc., and on the other hand life includes all the personal relationships, i.e., friends, family, self, neighborhood, etc. It is very important to focus on the level of work life balance among company employees because it is directly connected with the job performance. COVID-19 has adversely affected the world. The way of living, communicating, and traveling is also getting affected. All the stakeholders of society are rethinking over the previously learned strategies in order to overcome the downfall in every segment of economy. The new thoughtful ideas will help to support those who severely got affected in this crucial period (United States Statistical Division, 2021).

8.2 WHAT IS WORK LIFE BALANCE?

Why it is important to discuss about work life balance? Because the notion indicates that both the segments are related to everyone's life and second important issue is both the areas are very demanding and dominating in everyday life activities. In order to overcome the stress generated from everyday life, it is important to understand the harmonious way of having a smooth balance between the two. Work life balance is a strategy of distributing the time on the work or life activities in order to ensure the better accomplishment of their set targets and responsibilities. Work life balance indicates the smooth overlapping, division or sometimes imbalance between the parameters of personal life, i.e., family, friends, and self and the work, i.e., professional life including job and career. While the time distribution work is done it is more preferable to assign more time to the priorities in work as well as life. Here it is important to understand the difference between different terms like important, urgent, priority, etc. It is advisable to switch to those activities which receive more numbers on the above parameters that means those work which are really important, urgent, and are a great priority for the individual should be completed in time. The major focus on time management regarding work life balance indicates the better allocation of work life activities on the daily 24 hours time frame. The better and flexible the allocation is the chance are more to have better work life balance.

Work life balance indicates towards the strategy by which people manage the span of time spent at work and out of work areas. Outside work ideas are devoted to manage relationships building new relationship, fulfilling the family responsibilities carefully and effectively. The measure of tool which is being used to gain the optimal balance between the work and life is known as work life balance. Defining the work life balance in general is simple but there are many under cover challenges when it is talked about the different professionals personally.

It is one area where everyone is required to put forth their own personal efforts and their own new strategy to find out their own perfect way. The way by which we conceptualize the work life balance itself indicates the unique way to perform it (Defining Work Life, n.d.).

According to EY Global Generations (2015), as far as work life balance among employees is concerned with both the issues and strategies it is clearly related to their time management skills. Today time management skills work as a key to overcome the obstacles of work life balance. Work life balance was earlier talked in terms of female but now days it is gender neutral term. Work life balance is equally important in the world of business as it provides a platform to assess the problem areas of the job satisfaction level among the employees. A good work life balance is an indicator of balanced and smooth harmony among all the aspects of one's live. It is a learning process which is dynamic in nature (health direct, n.d.).

The purpose to define work life balance is to find out the purpose of work life balance positively as when we talk about the notion we simply try to understand the meaning and search the answers of why. The process of defining the same allows us to come up with the answers of some questions like why we need better income, why a new hose, why the kids, why to do a good job and so on (Work life balance, n. d.).

8.3 WORK LIFE BALANCE DURING COVID-19

During COVID like all other aspects of life, our 'Co-Life' concepts also changed their dimensions, and many new aspects are being introduced by this previously unknown situation of COVID era. The previously clear separation between work and life somewhere disappeared suddenly and the things started happening in haphazard way. We lost that harmony once but our struggle of gaining that harmonious level get started. We started rethinking, reevaluating the earlier effective strategies. No doubt this era brought some opportunities as well but the need of the ours is first to understand the notion and then learning to convert the newly emerged challenges in to the opportunities so as to reach to the optimal level of the work life balance.

While performing the job in the mode of work from life, it indicates towards the work culture where employees need not to come to the office but requires to perform their job tasks from home only. It provides the opportunity to the workers to enjoy flexibility as they can perform the work and job works easily altogether. The means of communications have added new features of comfort in this (Heery & Noon, 2017).

A research article published regarding benefits and limitations of work from home (2021) states that work from home (WFH) mechanism is welcomed in some sense as it provides the opportunity to work while staying at home; it eliminates the traditional way of work culture. It has totally reframed the concept of work, from the aspect of communication and the form of business style, but at the same time the recorded reduction in social interaction and standard norms of accountability should also be notices and critically understood.

By using the benefits of telecommunication employees get the opportunity to perform their job works while being at their own comfort places (Gajendran & Harrison, 2007). The term WFH has both positive and negative traits within itself. While maintain the COVID protocols it was sometimes required to work from remote areas, sometimes the workers had no choice for the same. This compulsory work is mostly full time instead of part time, but the positive aspect of the same is this research surveys indicated the positive work experiences of the employees (Cartmill, 2020; Galvez et al., 2020; Uresha, 2020). It has the advantage of flexibility, low cost for companies, motivation for work; more time can be devoted to fulfilling the family responsibilities, savings, etc. On another hand it has so many disadvantages also as while working from home so many employees felt the overburden of household duties, lack of personal space, overlapping of work and life, etc.

As discussed earlier work life balance is a gender neutral term now but some research evidences indicates that in comparison to men women are more stressed during maintain their work life balance as they feel more emotional disturbances and the lack of personal space during WFH periods. It is indicated that participative leadership can reduce the interference of work with personal life and then the employees can reduce their emotional exhaustion effectively (B., 2020). The mental conflicts were being experienced by the working women as there was a dual thinking process while maintain the perfection for both the work and the life while working as WFH (Delina & Raya, 2016).

8.4 REVIEW OF LITERATURE

8.4.1 AGE OF CHILDREN AND WORK LIFE BALANCE OF THE EMPLOYEES

Parenting is a very important and crucial social role. While having the responsibilities as a parent of younger children it is sometimes very difficult to manage between the different roles. May be due to the role conflicts the work life balance became a challenge. Early stages of parenting require lots of learning and efforts in order to provide a better parenting to the young ones.

Specially in the labor market employees it is a challenge to separate the work and life. A burnout state can lead to physical and mental exhaustion. It may be possible that the work ability can get adverse effect. Adverse effect of work life imbalance on the early stages of parenting was found.

The parents of preschool children are more prone to the mental exhaustion due to work life imbalance (Eurofound, 2017).

8.4.2 CONCEPTUAL DIFFERENCE AMONG DIFFERENT CONCEPTS RELATED TO WORK LIFE BALANCE

While explaining the term 'work life balance' there are different approaches which are used along with this such as work and private life balance, work, and family life equilibrium, etc. Researchers have indicated many dimensions in the field of work life balance. Earlier researches were focused to study the factors which were related for maintain the balance between work and life (Ramakrishnan, 2020).

8.4.3 THE SEGMENTS OF WORK AND NON-WORK LIFE

The awareness about work life balance itself initiates the more balance approach as simply the increased level of awareness provides the platform to judge the ineffective behavior patterns which rather intensifies the problematic behavior instead of proving the background for the smooth functioning.

The work life balance and the work personal life reconciliation are the popular terms which are used to enhance the awareness regarding work life smoothness. It is required to combine some areas and to reconcile some areas in order to have clear division among the work and non-work areas. The broader aspect of life includes the religious activities, community life, educational activities, and other related ones (Kopra, 2012).

8.4.4 CAUSES OF WORK LIFE IMBALANCE

The major cause of imbalance is the difference between the demands of work and life sometimes the work area is more demanding or sometimes the life one. If there is the equal priorities from the work and the life issues then the work life balance is perfect but when the problem is of imbalance then the notion is that the increased personal responsibilities, longer work hours, increased work commitments, additional responsibility of children along with the job performances. The positive orientation towards life is always a key to enhance the positivity, happiness, satisfaction in every segment of life. The researches have indicated that a positive work life balance is related to the effective reduction in the employees stress. The risk of burnout is reduced and the level of well-being is increased. The better aspect of this is that not only the employee but also the employer also get affected positively (Sanfilippo, 2020).

8.4.5 WORK LIFE BALANCE AND THE SOCIO-ECONOMIC SUSTAINABILITY

Work life balance has so many positive effects. The work life balance is the positively connected with the socio-economic sustainability. Higher levels of goals are related with the improved levels of work life balance. So many demographic challenges are linked with the work life balance issues. Lack of employment opportunities, low level of wages, gender difference among work life balances are the major issues which are related with work life balance. The need of work life balance changes with the change in our family and family pattern including the age of our children. In this way the pattern of work life balance is dynamic in nature (Thirion, 2016).

8.4.6 REMOTE WORKING AND THE WORK LIFE BALANCE

Remote working is the tool which provides the opportunity to the employees to fulfill the some of the life responsibilities while on the work hours. Early in the century remote working was explained as the important tool for the important tool for promoting work-life balance. Particularly the employees with children are the comfortable with the technique of remote working as it provides the opportunity to fulfill the variety of family responsibilities including the breast feeding, which is particularly useful for the employees with the young children, taking care of sick children, looking after of children during school holidays. The practice of remote working regularly additionally related with the reduction in work-related expenses such as traveling expenses. It also results in the saving of time spent on the way home (Hein, 2005).

8.4.7 WORK FROM HOME (WFH) DURING COVID-19 AND STRESS AMONG EMPLOYEES

Many things changed during the global pandemic due to COVID-19. A change was also observed in the working pattern and acceptable form of work culture. The new fashion of work also tried the WFH style. This change was not willing rather forced at the very onset. Initially when we were not prepared for work from culture but gradually the employees started perceiving the positive aspects of the WFH. It was perceived as the positive factor for the work life balance. However, as the situations prolonged due to the different waves of COVID the negative aspects start appearing and the employees started reporting WFH as the obstacles in work life balance. While working on the mode of WFH the connection between the employer and the employee was only at the distance of one communication by any means. Somehow the 24/7 job alertness and sudden job schedules started creating stress among employees. Hence the job hours restrictions were smartly removed, and the outside working hours expectations stepped in silently. At the same time always available interaction with family members is also not welcomed as the crossed boundaries interaction resulted in over interference in the work and life segments. The sudden emerged situations demanded the employees to try out the new strategies during the COVID pandemic, many mistakes were made. This new strategy experiment gave birth to new dimensions of work life balance. COVID has taught so many lessons to the mankind. The importance of cooperation is one of them. While having the trait of cooperation employees as well as employers gain a lot in terms of better work life balance.

Personal happiness, company retention and economic growth all revolves around the healthy work life balance. The work life balance also work as a motivator for the employees as the satisfied areas of their life energies the positive feeling and energy among them. This positive flow of energy ensures the better quality of work life balance (Ramakrishnan, 2020).

8.5 MAJOR CHALLENGES PERCEIVED

Lockdown period was not that easy. Everyone was in trouble, some because of financial loss, some due to emotional loss, some because of unexpected discomforting restricted movements. This strange situation somehow suffocated the personal space and extreme closeness among relatives living in the same house somewhat impacted negatively!

8.5.1 FAULTY COMMUNICATION STYLES

Communication is the main connecting agent which helps us to express our message in the form of feelings and thoughts to the other individuals. At the same time only by the means of communication we get connected with the entire world around us. There are many popular modes of communication including online and offline communication. We can get connected and communicate anyone with fraction of seconds. As we know the stimulus response formula of behavior which explains the phenomenon that our behavior acts as the stimulus to the other individuals who give their reactions or the responses in the form of their behavior. Sometimes due to the usage of faulty communication style we provide the platform for the other individual to misunderstand our message. The cycle go on and at last we reach to the point of unhealthy communication among different personal as well as professional relationships We have to take notice on the communication style which we use while interacting with our family members, whether we are aggressive in our communication or simple submissive. Post-traumatic stress disorder symptoms are associated with the stress provoking everyday life experiences and faulty communication styles (Acuna et al., 2017). During COVID pandemic it was observed that every individual got frustrated during lockdown period. Those who were unable to manage their emotions indulged in to aggressive communication styles. Being assertive can only help to smoothen the communication with our family members and at workplace. The anxiety due to potential risk of COVID-19 has also accelerated the risk of PTSD and traumas connected with stress. Better and healthy social relations are helpful in post-traumatic growth whereas lack of social connectedness is the big cause of continued negative and adverse health states. This is the global phenomena that status of social connection has a significant association with the PTSD symptoms (Matos et al., 2021).

8.5.2 UNREALISTIC DEMANDS

Fruits of expectations are bitter not always but many times. When we start expecting the things, favors, particular behavior from the other individual then unknowingly we employ the method of projection and in a way, we start controlling the behavior of others. In reality it is not possible to control the behavior of others but this unrealistic way of thinking provide us the idea to expect more and more from the other individuals. This leads to the vicious cycle of increasing levels of expectations from others.

Happiness is significantly connected with the positive expectations which one has from the future endeavors. The happiness or the positive outlook towards upcoming events filled us to start doing efforts for the accomplishment of the particular expected goals (Hagan, 2018).

Sometimes unknowingly we demand unrealistically form the family members and our colleagues. These unrealistic demands were on the geared up mode during the COVID pandemic. This contributed a lot in increasing the challenges of the family dynamics and work life balance. It is very much natural to expect from our loved ones. At one point of time from our frame of reference it is sometimes perfectly perfect but unknowingly and unintentionally these expectations are somewhere linked with the tendency of expecting more and more from others, which can ultimately affect the relations adversely. We know that it is not possible to have a perfect relationship. It is very much natural to have ups and downs in any relationship. It is always recommended to have a selfless relation (Bates, 2016).

Finally when we are not able to receive the expected behavior from the other individuals, unknowingly the frustration occurs who silently start nurturing the disguised aggression. These unfulfilled expectations can thus come out in different attires and can have reflections during our behavior with others.

8.5.3 PRACTICING TOXIC BEHAVIORS

It is the human nature that we need the immediate fulfillment of our desired behavior. The tendency to think ourselves as the well-wishers of others continuously motivate us to pursue others to correct their behaviors. At this point of time we sometimes forget to understand that the nagging is toxic habit as it hinders the individuality of others.

Toxic habit nagging is one of the challenge while maintain the family relationships as well as behavior with the colleagues. We keep on asking for the desired behavior from our family members, on the contrary we obtain only the failure in achieving the expected one. As during lockdown we came excessively closer with restricted movement so it became more stressful. While trying to make our relatives understand we simply jump in to criticism, in a way we ignore their efforts and simply instruct them for their failures. This criticism simply can't help to enhance the level of our family relations. During lockdown it was a challenge to make the family understand about the issues related to the job targets.

8.5.4 OVERLAPPING OF WORK AND LIFE

Work and life are the continuously overlapping and interfering domain of life. No one can ensure the equal distribution between the work and life segments. Both the notion can go together smoothly only when allowed to share time, energy, devotion, efforts with each other as well. It was a great challenge to simply segregate between the work and life both get mixed up with the newly emerged concepts of WFH, online classes, lockdown 1.0 and 2.0 processes, etc. That simply mixing up of all things and segment of life really initiated challenges and resulted into disturbed family relationships. It was a weird overlapping of personal and professional issues. For women and men alike, the working sphere and the private sphere are not entirely separate. There are intrusions in both directions: work encroaches on the domestic sphere and similarly, private life spills over into the workplace. This overlap sometimes leads to better organization but at the same time it has the potential traits of creating tense environment.

It is more common to see work affecting family life than the contrary. Men and women today are equally concerned by the intrusion of work into their private lives. However, women tend to be more preoccupied by their family lives at work than their counterparts, as they still assume the majority of domestic responsibilities. With the technological advancement and changed socially accepted norms, it is more likely to accept the concept of work life balance in gender neutral terms. Gone are those days when it was considered that the household work is the business of women only. The emergence of COVID no doubt increased the challenges for the work life balance where while practicing the WFH culture we are available both for the work and life responsibilities as well.

8.5.5 INABILITY TO PRIORITIZE DAILY CHORES

Prioritizing daily chores simply indicates the life skill where we can judge and arrange the day today life chores in a beautiful way. The practice of making a to-do list for work as well as life for the daily basis will definitely help out in the smooth functioning in both areas. With the advancement in the technology, we are enjoying its fruits in term of comfort but at the same time we are also in the easy grip of the development of procrastination tendency. Where we simply linger on the tasks and responsibilities and then feel ourselves overburdened with the heap of pending tasks and get stressed as the deadlines approaches.

The inability of prioritizing between the daily chores is one of the major problems of disturbed family relationships. Sometimes it happens when we simply feel overburdened and in hustle bustle manner we put the daily chores in a weird way.

During dealing with the time bound assignments many life skills are important to learn. One of these skills is the ability to differentiate between the urgent and important. It is always preferable to select the important tasks which can only be performed effectively and efficiently by the strategic management. This technique is useful for segregating and prioritizing among the heap of the emails, job assignments and all (Ghanagam, n.d.). It is important for the upliftment of the organizations to understand the notion that they should make their employees understand regarding their importance in that particular organization. The communication to the employees that they are really wanted and significant contributors for the well-being of the whole organization can fill up the employee with the positive energy and perception. It creates a motivated work environment (Kirloskar, n.d.).

8.5.6 CONFLICTS DUE TO CONTINUOUS SHIFTING FROM PERSONAL ROLES TO PROFESSIONAL ROLES

Recall the pre COVID days where we could simply justify our work hours for let's say '10.00 AM to 5.00 PM.' But in this new normal your normal office hours also changed or forcefully and informally converted into 24/7. This is because we are easily or have to be available every second for both the work and life tasks. The continuous shift from personal to professional roles was a major cause of conflict during pandemic situation. When you shifted from personal to professional roles it was not known. It created a lot of stress and challenging situation among working individuals. This continuous shift also has one more issue that is the uncertainty of this shift. The shift is suddenly demanded and the individual is expected to be there with perfection.

8.6 SUGGESTED STRATEGIES

8.6.1 COMMUNICATE RATHER GIVING INSTRUCTIONS

The communication is better to make the things done instead of the way of giving instructions always. At the workplace as well as at home it is better to ask for the involvement from the coworkers as well as the family members. While working together it is advisable to share the responsibilities as well as the success with the team members. At work place we have the coworkers and at home we have our family members. Whenever the habit of giving instructions is practiced, we simply encourage the tendency to avoid or simply ignore the right of others to express their shortcomings, likes, dislikes, comfort zones, strengths weakness and so on.

8.6.2 PRACTICE TONIC HABITS INSTEAD TOXIC HABITS

Practicing the habit of tonic habits like acceptance, empathy, and positive regards can help instead of practicing toxic habits like nagging, anger, blaming, criticism, etc., it would be better to take a shift from toxic to tonic habits. Positive change is also required and welcomed in the work as well as life settings. While practicing the toxic habits we sometimes fail to understand that we can stop anytime as soon as we get the clarity about our negative form of behavior and we can instantly start practicing the fruitful ones. There is no requirement of the major efforts. A simple smile, a short silence can do wonders. Like learning, relearning, and unlearning are equally important.

8.6.3 REACH OUT TO FAMILY MEMBERS AND COWORKERS

Try to sit with the family members and also with the colleagues and give time to listen their requirements. Assisting their mental health needs and giving quality time to the individual will definitely help to overcome the communication gap. This is the irony of the hour that we are losing the connectivity in real world social relations while maintain the relationships on social sites, while enjoying the benefits of virtual reality we are losing the grip from physical reality. It would be of great help if we start asking for favors and seek help from our coworkers and the family members.

8.6.4 BE ASSERTIVE, EXPLAIN YOUR LIMITS

Being assertive means standing for one's own rights. It is in no way a fight in verbal form but simply a strong assertion of one's own needs for the sake of their rights. During the process of socialization, we learn many acceptable ways of behaving in our social settings. May be because of such observational learning we have leant some easily acceptable communication styles which we learned as great but I reality they may not of that much use for us. The practice of being assertive and the ability to communicate that you also have your limits can help in the smooth communication from the family members and coworkers. Sometimes due to our inability to express our motives we give rooms for the misunderstandings. When the message is not explained properly it can invite hostile relationships.

While practicing the communication style of assertiveness there is less rooms for the miscommunication and at the same time one is able to stand of the rights of own and the other person. In this way they one can be in the state where he/she less number of conflicts with others (Ames, 2017). Assertiveness can be an important agent for the initiation and maintenance of the self-esteem of any individual. In this way one can achieve the goals of higher levels of self-esteem (Unal, 2017).

8.6.5 CONTRIBUTE TO MAKE ENVIRONMENT POSITIVE

The term environment has many fold explanations. In terms of behavior, we understand the environment as the region from which the stimulus is received it may be internal environment or external environment. In terms of communicating with others we are talking about external environment. In terms of mental processes our focus is on the internal environment. The environmental factors can initiate the irritation, frustration, anger, anxiety, fear or any more negative ones. But our responsibility as a positive individual should be to contribute in such a way that it can enhance the positivity in the environment not the negative one. When we simply accept the thought of others with the positive expression even we are not agreeing with the same, can provide us an opportunity to explain our disagreement very politely to the other.

Better to contribute to make the family and work environment positive rather than to make it more hostile and problematic. Sometimes unknowingly we make our habit to practice those faulty communication styles which contributes in making the environment more negative and hostile. On the contrary one should practice the tendency of 'let go' so that a new positivity can be felt among the family members and coworkers.

8.6.6 BE MINDFUL OF THE PRESENT

The best way of living is to enjoy the present moments of the life, as we are not aware of the things that what is going to happen in the future and at the same time, we have no control over the already occurred events of the past. Then why not we enjoy the present moment which is provided by Almighty with lots of new energy for us. Being mindful for the present moment is the virtue which can only be gained by the increased levels of awareness about our surroundings. One can practice this by using simple techniques like grounding techniques which indicates to have a sense of touch with the land barefoot preferably over the green grass, if not possible it can also be practiced simply on the floors available. The trait of detachment from the environment can provide a sense of relaxation; a 2 minute break can help to quickly come out of the stressful environment.

Try to be in the present situation instead of jumping into the negative expectations of the future. The tendency of negative outlook of future can hinder the well-being of the individual as well the family as whole. Being mindful about the present will help to understand the accurate challenges and opportunities, while negative outlook can force to indulge into the unrealistic anxiety feeling.

While being mentally present in the current moment, we have the chance to enjoy the happiness of the moment as every new second gives us a new life to live. With such a positive attitude and acceptance of the reality can definitely provide us abundance of positivity around us. Our mental traveling between negative and positive life events in the range of past and future expectations tend to provide the risk of being caught in the net of negative mental states. It is of no use to live our life only in the expectations of the future. The future expectations are filled with uncertainty; hence it can trigger the anxiety for the upcoming life events (Mindfulness, n.d.).

8.6.7 BELIEVE IN YOURSELF AND HAVE PROPER BREAKS AND FUN ACTIVITIES

The concept of individual differences advocates that every individual is unique and cannot be replaced by the other one. When we unique and definitely have certain qualities then why we don't love ourselves. The issue is we don't know ourselves. This is the plight of the hour that we are not even know ourselves and rather we are not interested in the same. Meeting with self-techniques like self-talk is important. But here the caution is negative self-talk can push you down so it is important to understand the practice of indulging into positive self-talk is should be encouraged. Another good strategy is journaling your behavior which will help to understand the hidden notions of your behavior. The tendency to have question on one's own abilities and skills is the major issue. We sometimes take notice only on the shortcomings of ourselves, our employers, coworkers as well as of other family members. This habit can result into the adverse results in the form of lack of confidence, lack of interest, irritation, depression, and many more. On should have proper breaks and fun activities as a platform to celebrate the small moments of life. This optimism will definitely help in the upliftment of society well-being in general and work life well-being in particular.

8.7 CONCLUSION

The COVID era was new but not normal. The journey from new to normal was not that easy and smooth. All of us had to go through the mental and emotional turmoil and many ups and downs. Our family relations and professional relationships were also got affected by them. There were newly emerged dimensions. To secure an uninterrupted and secured work and life activities, it is required to define the limits of both the personal and professional matters. This contour framing will serve as the basis for the mutually separated but interrelated performance for the co-life issues (Clutterbuck, 2003). A habit of following the work schedules and the prioritized life chores help us to feel a sense of well managed living, but if at the same time we are not able to understand the nature of work life demands then it can cause the sense of question over the self-efficiency. It is important to have a flexible work life hierarchy for everyday life (Jefferson Center, 2021).

Lockdown period acted as an opportunity to think over the challenges being experienced and to explore the new effective strategies to deal. Even the small initiative regarding addressing and changing the toxic habits into the tonic can do wonders. Though these challenges were there in existence in our lives earlier in pre COVID era also but the COVID era has introduced more complexities in our simple daily life. Who could thought that being at home 24×7 can be a source of stress for us! While being social creatures we have to practice the social distancing, we human being the highest creature of God will have to remain in lockdowns. This COVID has brought so many unexpected changes in our lives. The new strategies were required to deal with the new issues. Hence it can be concluded that each and every individual of the family at home and also the coworkers at workplace has to contribute to make our work life balance more happy and positive. Only work life balance is the one which bring health and harmony for the employee and at the same time for the employer. This balance beautifully invites the well-being among all the different dimensions of one's life. The ultimate objective is to be happy that can be initiated with the focused efforts of the balance co-life.

KEYWORDS

- **COVID**
- **effective strategies**
- **ineffective strategies**
- **new normal challenges**
- **remote working**
- **socio-economic sustainability**
- **work life balance**

REFERENCES

Alejandra, A., & Sheryl, K., (2017). Family communication styles and resilience among adolescents. *Social Work, 62*(3), 261–269. https://doi.org/10.1093/sw/swx017.

Ames, D., Lee, A., & Wazlawek, A., (2017). Interpersonal assertiveness: Inside the balancing act. *Soc. Personal. Psy. Compass, 11*(6). doi: 10.1111/spc3.12317.

Bates, T., (2016). *Unrealistic Expectations and Relationships: 5 Key Signs*. Retrieved from: http://www. https://psychcentral.com/blog/relationship-corner/2016/09/unrealistic-expectations-and-relationships-5-key-signs#1 (accessed on 7 February 2023).

Bhumika, (2020). Challenges for work–life balance during COVID-19 induced nationwide lockdown: Exploring gender difference in emotional exhaustion in the Indian setting. *Gender in Management, 35*(7, 8), 705–718. https://doi.org/10.1108/GM-06-2020-0163.

Clutterbuck, D., (2003). Managing the work-life balance. *Journal of Management,* 101–120.

Defining Work Life Balance: Energy is the Missing Ingredient, (n.d), Retrieved from: http://www.kumanu.com/defining.worklifebalance_its_history and future (accessed on 7 February 2023).

Delina, G., & Raya, R. P., (2016). Dilemma of work-life balance in dual-career couples-a study from the Indian perspective. *International Journal of Indian Culture and Business Management, 12*(1), 1–27.

Eurofound, (2017). *Working Time Patterns for Sustainable Work.* Available online at: https://www.eurofound.europa.eu/sites/default/files/ef_publication/field_ef_document/ef1720en.pdf (accessed on 7 February 2023).

Gajendran, R. S., & Harrison, D. A., (2007). The good, the bad, and the unknown about telecommuting: Meta-analysis of psychological mediators and individual consequences. *Journal of Applied Psychology, 92*(6), 1524–1541.

Gálvez, A., Tirado, F., & Jesús, M. M., (2020). Work–life balance, organizations and social sustainability: Analyzing female telework in Spain. *Sustainability, 12*, 3567. 10.3390/su12093567.

Global Generations, (2015). Retrieved from: http://www.survey.ey.com (accessed on 7 February 2023).

Hagan, (2018). *Psychology of Expectations.* Retrieved from: http://www.psychologytoday.com (accessed on 7 February 2023).

Hein, C., (2005). *Reconciling Work and Family Responsibilities: Practical Ideas from Global Experience.* Available online at: https://www.ilo.org/public/libdoc/ilo/2005/105B09_142_engl.pdf (accessed on 7 February 2023).

Jefferson Center, (2021). Retrieved from: https://www.jcmh.org/how-to-maintain-a-work-life-balance-during-coronavirus/ (accessed on 7 February 2023).

Korpa, V., (2012). *Darba un Ģimenes Dzīves Saskaņošana Privātā Sektora Organizācijās (Reconciliation of Work and Family Life in Private Sector Organizations).* PhD thesis. Available online at: https://www.szf.lu.lv/fileadmin/user_upload/szf_faili/Petnieciba/promocijas_darbi/Viola_Korpa_2012.pdf (accessed on 7 February 2023).

Matos, M., McEwan, K., Kanovský, M., Halamová, J., Steindl, S. R., Ferreira, N., Linharelhos, M., et al., (2021). The role of social connection on the experience of

COVID-19 related post-traumatic growth and stress. *PLoS One, 16*(12), e0261384. doi: 10.1371/journal.pone.0261384. PMID: 34910779; PMCID: PMC8673633.

Mindful: Healthy Mind, Healthy Life, (n.d.). Retrieved from: http://www.mindful.org (accessed on 7 February 2023).

Parent-Thirion, A., (2016). *Working Hours are Decreasing, So Does Work–Life Balance Still Matter?*. Available online at: https://www.eurofound.europa.eu/sites/default/files/ef_publication/field_ef_document/ef1649en.pdf (accessed on 7 February 2023).

Ramakrishnan, R., (2020a). Work life balance during lockdown. *SSRN Electr. J.*, 1–12. doi: 10.2139/ssrn.3717067.

Ramakrishnan, R., (2020b). Work life balance post COVID-19. *SSRN Electr. J.*, 1–17. doi: 10.2139/ssrn.3717081.

Sanfilippo, M., (2020). *How to Improve Your Work-Life Balance Today.* Available online at: https://www.businessnewsdaily.com/5244-improve-work-life-balance-today.html (accessed on 7 February 2023).

The Pros and Cons of Work from Home, (2021). Retrieved from: https://www.indeed.com/career-advice/finding-a-job/the-pros-and-cons-of-working-from-home (accessed on 7 February 2023).

Toyota Kirloskar Motor, (2021). Retrieved from: http://www.timesofindia.com (accessed on 7 February 2023).

Unal, S., (2012). Evaluating the effect of self-awareness and communication techniques on nurses' assertiveness and self-esteem. *Contemp. Nurse, 43*(1), 90–98. doi: 10.5172/conu.2012.43.1.90.

United States Statistical Division, (2021). *Report on the COVID-19 Impact on Public and Private Life as Well as Statistical Work.* Retrieved from: http://uis.unesco.org/en/news/report-covid-19-impact-public-and-private-life-well-statistical-work (accessed on 7 February 2023).

Work Life Balance, (n.d.). Retrieved from: http://www.worklifebalance.com (accessed on 7 February 2023).

Work Life Balance. Health Direct. Retrieved from: https://www.healthdirect.gov.au/work-life-balance (accessed on 7 February 2023).

CHAPTER 9

Effect of Autogenic Relaxation and Self-Management Training on Internet Addiction Among Red Collar Employees in Post-COVID-19

NEHARSHI SRIVASTAVA[1] and NEETA GUPTA[2]

[1]*Assistant Professor, Amity Institute of Behavioral and Allied Sciences, Amity University, Jaipur, Rajasthan, India*

[2]*Associate Professor, DAVPG College, Dehradun, Uttarakhand, India*

ABSTRACT

The coronavirus infection 2019 (COVID-19) flare-up has universally disturbed daily routine. Amid this outbreak, individuals worldwide were anticipated to come across mental health challenges. In specific, internet addiction may have ended up as a genuine issue among all the populace and especially for red collar employees. However, very few studies have been conducted on internet addiction among red collar employees during COVID-19 so, this study aimed to examine the effect of group therapies (ART and self-management training (SMT)) for reducing internet addiction among Red Collar male and female employees after the pandemic. For this purpose, a total of 200 respondents from different in Uttarakhand were contacted and selected using Quota Sampling, who were further equally divided into males and females randomly into equal groups exposed to different types of group therapy. Internet addiction test (IAT) constructed by Young et al. (1998) was used to measure Internet Addiction of the

Reshaping the Business World Post-COVID-19: Management Strategies for Sustainable Behavior Change. Arvind K. Birdie & Ruchi Joshi, (Eds.)

adults. A 2×2×2 factorial mixed design was utilized to conceptualize the study and analyze the gotten information. After information collection, pre-test, and post-test of ART and SMT were analyzed by utilizing SPSS.20 software. The gotten information were analyzed utilizing Implies, SD, and ANOVAs. Internet addiction was found high among red collar employees after the pandemic (pre-test). The post-test results have suggested that SMT was found to be a better technique for reducing Internet Addiction red collar employees. Females were found to report more internet addiction than males. Results from this study can be used to help mental health organizations and educational agencies design programs that will help p Internet addiction prevent in young employees after COVID-19 pandemic.

9.1 INTRODUCTION

Since December 2019, COVID-19 has risen and spread quickly over the globe. The World Health Organization announced COVID-19 widespread on March 11, 2020. Due to concerns approximately disease and shame toward tainted people and their families, as well as word related components, economy, segregation, forlornness, and other components, the predominance of different mental disorders such as Anxiety, depression, stress, and insomnia has expanded. To manage with these modified circumstances, individuals are more likely to enjoy in addictive behaviors (counting substance utilization and behavioral addictions) and/or backslide from restraint. Amid the widespread, to moderate down the spread of the infection, individuals were prompted to remain at domestic, and open-air social exercises were definitely decreased. Observational prove proposes expanded consumption of online amusement. An internet study conducted by Angang Organization (2020) on the common populace in China 46.8% of participants reported increased dependence on Internet use, while 16.6% reported longer Internet use duration. Center for Japanese Mental Health (2021) conducted a gender-based study on Internet addiction. Results appeared that males regularly confine themselves from the social community in arrange to engage in online gaming whereas females utilize the internet as to not be prohibited from their communications online. In spite of the fact that the stay-at-home degree and Internet utilization might encourage the anticipation of further contamination spread and control the widespread abuse of the Internet and gaming can lead to behavioral enslavement. Psychological distress has ended up a critical public health issue amid the pandemic and has pulled in the attention of analysts and the open alike. Now a days Social media has ended up versatile and open and websites are utilized in day to day communication for the reason of sharing common thoughts and considerations, for sending and accepting diverse sorts of data based on individual or official work perspectives.

Social media clients ordinarily get to the administrations of social media through the internet or other internet-based innovation on their portable workstations or desktop computers or by downloading applications that amplify the usefulness of social media and social systems to portable gadgets, such as smartphones or tablets. Multi-national companies have an incredible fear that organization might get influenced antagonistically since representatives are paying more consideration towards social media locales. Innovation advancement made the learning simple for everybody. Improvement of shrewd phones, web openness and different applications encouraged individuals to memorize anyplace and anytime. Social media has gotten to be a necessary portion of most people's lives.

Social network service is additionally known as SNS. A SNS is an online service stage or location that centers on encouraging the building of social systems or social relations among individuals who, for illustration, share interface, exercises, foundations, or real-life associations. SNSs such as Facebook, Twitter works like this. Individuals who gotten to be individuals of these SNSs communities, firstly get an individual profile that shows their individual arrangement counting the title, work, photo, relationship status, religion, leisure activities and so on. At that point systems of companions are made and other clients can at that point press on their profiles and navigate ever broadening social networks. In the last few decades, there has been a tremendous advance in the field of information technology and communication, especially during COVID-19. This transformation has happened as a result of the presentation of the internet with which the world has been seeing extraordinary changes in nearly all angles of our cutting edge life. Internet has empowered the productive utilization of innovative gadgets such as Smartphones, iPods, and other internet-devices to be displayed inside the work environment. This nearness can have both positive and negative affect on the efficiency of red collar workers. The negative effect is appeared when red collar employees tend to grasp exercises that can devour the company's profitable time. These exercises incorporate browsing the internet, checking social network websites, texting, and online chatting with companions. Such exercises decrease the organizational yields since of destitute worker execution as the profitable time is squandered coming about in negative results upon the full efficiency of the commerce organization or company. In addition, it was found that 80% of the companies communicated that their employees had abused the internet in their work environments by engaging in non-work-related exercises on the internet amid the work hours. That is why some companies have resorted to using job-specific tactics to avoid loss of work productivity resulting from the usage of the internet in the workplace. A few of these strategies incorporate blocking certain internet destinations, restricting the uses of smartphones and observing the internet utilization inside the workplace. Industry examiners appraise misplaced income worth billions of dollars, crediting it to representative Web mishandle. Patterns moreover recommend that misplaced work efficiency and corporate obligation have risen as modern working environment concerns due to development of modern online advances and portable computing. Such worker Web abuse makes unused administration predicaments on how to reply to frequencies of such abuse as well postures organize security dangers and channels on organize transmission capacity. Inside an organization, it is basic to guarantee that workers are utilizing computing assets viably and fittingly. A large Number of Studies on internet uses were focused on the Impact of Mental Health while little is known about the impact on employment productivity during the COVID-19 lockdown. Many studies indicate that red collar employee internet abuse at work-place leads to loss of productivity, negative publicity, and potential legal liability. In a survey of human resource directors carried out by the Society of Human Resource Managers, (2019). It was found out that nearly 70% of the companies surveyed provide internet access to more than 50% of their working staff. Clearly, the excessive use of the red collar employees' smartphones at work place proved to affect work productivity negatively. According to Fludd (2014), a huge number of red collar workers appear to enjoy in internet exercises that devour the company's beneficial time. These internet exercises demonstrated to be efficiency adversaries as they lead to destitute execution of representatives who were actively utilizing the internet amid work hours and discover troubles in accomplishing their perfect efficiency levels Indeed more awful, others come up short to reply to customers' in an opportune way uncovering client fulfillment to be misplaced, hence eating into the company's benefits The discoveries compared with the comes about of Grodzinsky & Gumbus (2005) consider which found that 80% of the companies included within the test conceded that their workers had abused the internet through carrying out non-work related exercises at working environment. That's why commerce supervisors are anxious of the destructive impacts of utilizing the web at work-put that breach the routine standards of the trade organization, debilitating its well-being. The coronavirus disease 2019 (COVID-19) outbreak has significantly disrupted normal activities globally. Amid this epidemic, individuals around the world were anticipated to come across a few mental health challenges. In specific, Internet habit may gotten to be a serious issue among all the populace and particularly for red collar workers (Srivastava et al., 2021). The term employee internet abuse is characterized as "any purposefulness act by laborers to utilize the web for non-work-related exercises inside the official working hours" (Qinyu, Xin, Anil, & Long, 2009). A disjoin illustration illustrating the issue of internet addiction and abuse at work environment is the situation in which three committee workers in Wales misplaced their employments since of their intemperate time they went through online on the site of eBay sell off whereas they were at the working environment. Their bosses were moreover considered capable for their waywardness since of permitting the three representatives to utilize the internet whereas they were at work put thus (Qinyu, 2009).

According to the Bangladesh Telecommunication Regulatory Commission, there were 99,428 million (approx. 60.9% of the populace) internet users in February 2020. In spite of the fact that there's recounted early prove that this may lead to expanded efficiency, it has moreover driven to expanded push where workers must learn modern innovations, be accessible for work at nearly all times, remain with computerized gadgets all the time, and adapt with multi-tasking. The utilize of video-conferencing and substance conveyance administrations such as Zoom and Akamai too expanded. Those working from domestic utilizing this innovation have found themselves beneath strongly examination and tension. That a huge portion of the people's social and passionate development is happening whereas on the internet and cell phones may influence the physical health as well as mental health of people considering the long hours went through on the internet amid the obligatory stay-at-home orders, which uncovered the populace to parcels of deception known to influence physical and mental health.

The study has tried to explore the effect of group therapies (autogenic relaxation training and self-management training (SMT) for reducing internet addiction among Red Collar male and female employees after the pandemic.

Relaxation can offer assistance to diminish the indications of stress and anxiety. In spite of the fact that the cause of the anxiety will not vanish, an person will likely encounter upgraded capacity to bargain with it once the pressure within the body is discharged and contemplations gotten to be more clear. The autogenic relaxation technique involves relaxing the muscles to create one feel calmer.

Autogenic implies something that comes from inside. In this relaxation technique, a person employments visual symbolism and body mindfulness to diminish stress. One rehash words or proposals in his/her intellect to relax and diminish muscle pressure. For case, an person may envision a tranquil setting and after that center on controlled, relaxing breathing, abating his/her heart rate, or feeling diverse physical sensations, such as unwinding a body part.

Self-management skills have been defined as the abilities for adaptive and positive behavior that enable individuals to deal effectively with the demands and challenges of everyday life (WHO, 2019). Adaptive' means that a person is flexible in approach and is able to adjust in different circumstances. Positive behavior' implies that a person is forward looking and even in difficult situations, can find a ray of hope and opportunities to find solutions. Self-Management skills include psychosocial competencies and interpersonal skills that help people make informed decisions, solve problems, think critically and creatively, communicate effectively, build healthy relationships, empathize with others, and cope with managing their lives in a healthy and productive manner.

9.2 WHY GROUP ART AND SMT?

Group therapy could be a kind of psychological therapy that takes place with a bunch of individuals together instead of with a person amid a one-to-one session. Whereas the term can actually be connected to any kind of psychotherapy that's delivered to a bunch, it is most commonly related with a particular treatment sort that produces utilize of the bunch energetic. Gather treatment can be based on any hypothetical approach, from cognitive-behavioral to humanistic. The present study has utilized two types of Group Therapy for intervention of internet addiction. They are autogenic relaxation and SMT. Group Therapy was specially planned for the present study because the study has tried to deal with a serious issue of Internet addiction of red collar employees,' so it was basic to induce them to realize that they are not the as it were ones who are confronting negative considerations. This feeling of universality makes the respondents relaxed and more comfortable in handling their problems. Another reason for using Group Therapy in the present study was that it took less time as attending to all the respondents individually in one to one counseling sessions would have proved very time consuming it.

9.3 REVIEW OF LITERATURE

Mukhtar et al. (2021) in their study pointed to look at the impact of internet addiction in the work environment on the execution of workers. Such internet habit in the workplace has demonstrated to be a later vital wonder. That's why numerous companies these days attempt to actualize certain viable measures and directions to avoid this web usage within the working environment to guarantee the good execution of their employees and to realize client fulfillment in conjunction with the great quality of the benefit given. \Findings of the study suggested that the internet addiction had a high effect on employee's productivity, and the results of the study were clearly demonstrative of the truth that the larger part spend more than 6 hours on the internet looking for entertainment.

Tanvir Abir et al. (2021) in their study inspected the effect of internet utilize recurrence on self-reported physical wellbeing amid the COVID-19 lockdown in Bangladesh. A web-based cross-sectional consider on 3,242 government representatives was conducted for a period of three months. The overview secured socioeconomics, internet use recurrence and self-reported physical wellbeing questions. Visit and broad employments of the internet were solid indicators of physical wellbeing issues, and the discoveries proposed the need for expanded mindfulness approximately the physical wellbeing issues that will get activated by over-the-top internet usage.

Yeonsoo Kim et al. (2020) to look at way of life patterns and dietary behavior based on the level of internet enslavement of Korean regulatory staff. Information was collected from 853 Korean regulatory staff. The level of Internet habit was decided based on the Korean Internet addiction self-scale short form for youth. The associations between the staff's levels of Internet addiction and way of life designs and dietary behavior were analyzed employing a chi-square test. Sporadic sleep times and the uses of liquor and tobacco were higher in high-risk internet users than no hazard internet users. Besides, in high-risk internet clients, unpredictable dietary behavior due to the misfortune of craving, a tall recurrence of skipping dinners, and snacking might cause awkward nature in wholesome admissions. Eat less quality in high-risk internet clients was too more awful than in potential-risk internet users and no chance Web users. Comes about outlined that high-risk internet users have unseemly dietary behavior and poor diet quality.

Weiss et al. (2019) that aimed to identify the degree to which gender influences suicide risk among employees with Depression. Study demonstrates that suicide risk is greatest for male and female during excessive use of the internet.

Jorm et al. (2019) examined the effect of Mindfulness Breathing on the level of depression and excessive use of internet among red collar employees. Results have shown that there is a significant effect of regular practicing of mindfulness breathing on internet addiction and depression level among red collar employees.

Henary et al. (2018) in their study examined the effectiveness of motivational therapy on decreasing stress, internet addiction and suicidal ideation of red collar female employees. One week earlier to last examinations, members were arbitrarily allotted to either motivational group or a control group. Stress Inventory was administered pre- and post-level. Motivational Therapy activities included motivational stories audio-video material and role playing. Result revealed that a brief period of motivational counseling significantly reduce stress, Internet addiction, and suicidal ideation, which may have implications for motivational therapy programs that offer strategies for making a difference red collar workers with internet addiction, stress, and suicidal ideation.

Hermenau et al. (2018) in their study aimed to investigate the effects of a cognitive behavioral intervention combined with a Progressive Muscular Relaxation Technique on stress among employees. This study supported the hypothesis that the cognitive behavioral intervention combined with a Progressive Muscular Relaxation Technique could decrease the stress level and internet addiction and the effectiveness was lasting.

Sharma et al. (2018) in a longitudinal study of over 300 patients with internet addiction found that women with higher severity ratings for internet addiction were at greater risk for suicide attempts, compared to men.

9.4 SIGNIFICANCE OF STUDY

Internet Addiction has become very common in the present scenario as due to Corona Pandemic use of internet has now become inevitable, and most of us are not even aware when this necessity becomes addiction. Now due to pandemic when the whole world is forced to live in the virtual world rather than the real world, the use of internet has started interrupting in the normal functioning of our day to day life and we are not able to survive without internet. Excessive internet use is increasing in every age group whether they are school going children, adolescents, adult employees, red collar workers or elderly people because we all are left with no other option except to communicate or get their work done through internet during this pandemic situation.

Excessive use brings with it unfavorable academic, social, and psychological consequences as well as the risk of internet addiction that is the situation in which the technology becomes bane. Nearly 90% of red collar workers are physically cut off from their work settings and are doing work from home (WFH) due to the COVID-19 pandemic, and over use of technology has become a necessity for them to access their clients and target group and to interact with each other.

The employment scenario world over has been significantly influenced by the widespread. In expansion to the danger to open wellbeing, the financial and social disturbance debilitates the long-term jobs and prosperity of millions. Nearly overnight, the Coronavirus (COVID-19) widespread has changed the work and working environments of the open and private division. Open hirelings are playing a driving part within the controlling the disease and overseeing different viewpoints of administration during the pandemic.

Healthcare workers are keeping restorative frameworks working and families secure. Governments over the world are attempting to find novel ways to plan and channel exceptional financial jolt investing and oversee serious spikes in unemployment. In nut shell it can be said that all the red collar laborers are compelled to utilize web for their work and got to be inclined to ended up internet dependent as working from domestic for benefit segment specialists and administrators gets to be the new normal.

Internet services have seen rises in usage from 40% to 100%. Weiss et al. (2019) compared to pre-lockdown levels. Video-conferencing administrations like Zoom have seen a 10 times increment in utilization, and substance conveyance administrations like Akamai have seen a 30% increment in substance utilization, once more signaling the internet as a need for red collar workers.

The impact of internet addiction should be further explored as it has the potential of causing serious damage to physical, psychological, and social lives of red collar employees. Furthermore, there is very little empirical literature or intervention for same related to the implications of internet addiction on red collar employees, the present study therefore was conducted to fill this research gap. Indeed in spite of the fact that social media has demonstrated to play a positive part in day to day human life for the reason of sharing and communicating with one another, it too has its share of negative encounters, and government organizations and representatives are no different. Most of employees are killing their time and organization time, browsing internet for amusement, at times indeed finding themselves in a humiliating circumstance. It was found that representatives spend most of their time on internet, they do pay consideration to their designated which in turn leads to client disappointment and maligning of the organization, in this manner coming about in misfortune of incomes, wastage of time and complaints by clients. The present study tries to assess and investigate the impacts of internet addiction among red collar employees and its management through group therapies.

9.5 RESEARCH METHODOLOGY

9.5.1 OBJECTIVES

- To assess internet addiction in relation to group therapies (autogenic relaxation training and self-management training) among male and female red collar employees.
- To explore gender differences on internet addiction among red collar employees.

- To explore interactions of group therapy (ART and SMT), conditions (pre and post) and gender (male and female) on all the dimensions of internet addiction among male and female red collar employees.

9.5.2 HYPOTHESES

- The respondents with autogenic relaxation training and self-management training will differ significantly on internet addiction.
- There will be significant gender differences on internet addiction among male and female red collar employees.
- There will be a significant difference in pre-test and post-test conditions of both the therapies (ART and SMT) on internet addiction among male and female red collar employees.
- There will be a significant interaction between gender (male and female) and therapy (ART and SMT) on internet addiction.
- There will be a significant interaction between conditions (pre and post) and gender (male and female) on internet addiction.
- There will be a significant interaction between conditions (pre and post) and therapy (ART and SMT) on internet addiction among male and female red collar employees.
- There will be a significant interaction between conditions (pre and post), gender (male and female) and therapy (ART and SMT) on internet addiction.

9.5.3 RESEARCH DESIGN

A 2×2×2 factorial mixed design was employed to conceptualize the study and analyze the obtained data. Three classificatory variables were used to classify the respondents into eight categories. The first classificatory variable being Group Therapy which was divided into two types, i.e., autogenic relaxation training and SMT which were matched on two conditions of therapy, i.e., pre- and post- (the second classificatory variable). The third classificatory variable corresponding to the sex of the respondents led to two categories of gender, i.e., males and females. The classificatory scheme yielding the respondents is as in Figure 9.1.

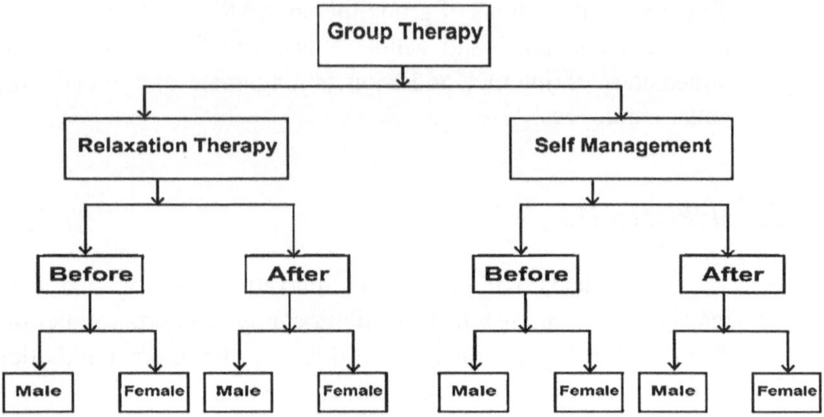

FIGURE 9.1 The distribution of the respondents.

9.5.4 SAMPLE AND SAMPLING

The purpose of the present study was to see the impact of autogenic relaxation training and SMT (types of group therapy) on internet addiction among red collar employees. For this, different government offices in Uttarakhand were contacted and the respondents were selected using Quota Sampling. A total no of 200 respondents were chosen through Quota Sampling from diverse government workplaces of U.K. and they were similarly divided into males and females. They were assist separated randomly into equal groups having distinctive sorts of group therapy. The sample scheme is given as in Figure 9.2.

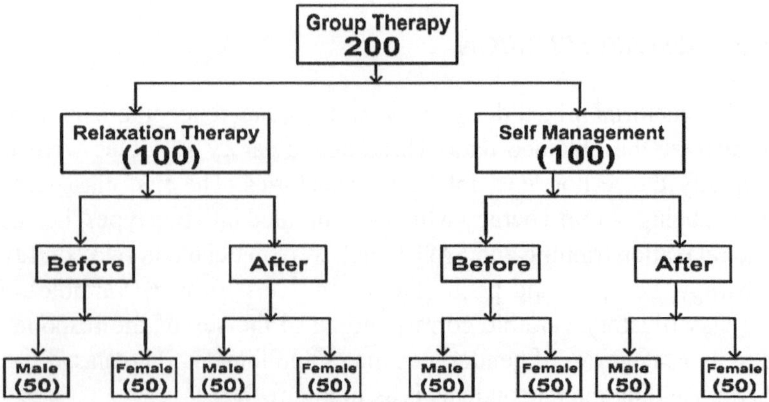

FIGURE 9.2 The distribution of the respondents.

1. **Tools Used:**
   - Internet addiction test (IAT) constructed by Young et al. (1998) was used to measure internet addiction of the adults.
2. **Therapies Used:**
   i. **Autogenic Relaxation Training:** Autogenic training could be a relaxation strategy to begin with, presented by German therapist Johannes Heinrich Schultz in 1932. To begin with, the learner is energized to settle into a comfortable position sitting upright, leaning back, or lying down. What is most vital is that the chosen pose advances calm relaxation. The therapist starts by utilizing verbal prompts to direct the trainee's breathing and coordinate consideration to certain parts of the trainee's body. The verbal prompts can energize sensations of greatness and warmth, which can at that point lead to profound unwinding. The specialist might lead the signals, have the learner rehash them, or have the learner say them noiselessly, depending on the level of preparing the person has received.

   Some of the verbal cues that may be learned include (Cuncic, 2020):

   - I am totally calm (say once);
   - My right arm is overwhelming (say six times);
   - My right arm is warm (say six times);
   - I am totally calm (say once);
   - My heart beats calmly and frequently (six times);
   - My breathing is calm and standard…;
   - It breathes me (six times);

   Once the session is wrapped up, the specialist will offer assistance to the learner "cancels" the unwinding session. One state that's commonly utilized is "Arms firm—Breathe deeply—Open eyes." The session concludes, and the learner is energized to practice what has been learned at home, outside of the session. Each session extends on the final lesson until the learner and specialist feel confident that the method can be conducted independently.

   ii. **Self-Management Training (SMT):** It is a systematic intervention approach for evoking change in among employees.

It is based on principles of motivational psychology and is designed to produce rapid, internally motivated change. Self-Management practices direct a person on the path of happiness, compassion, selflessness, change negative thoughts into positive thinking (which help a person to cope with depression and suicidal ideation) and improve positive coping skills. There are many Self-Management techniques in tradition which could play an important role in making life better and which may have therapeutic effects. This SMT is the processes of realization about the qualities of the individual which differentiate him from others. SMT was given by the SMT manual (Gupta & Neharshi, 2018). This manual consists of various experiential and motivational activities:

a. **Self-Awareness:** It is an experiential learning used in the present study which is very close to Self-Actualization (Concept of Maslow).
b. **Effective Communication Skills:** These are very important because it helps to improve teamwork, decision making and problem solving approaches.
c. **Motivation:** It is also experiential learning which helps to remove the negative thoughts and creates positive thinking. It is a combination of humanistic treatment and enhance cognitive behavioral strategies designed to treat depression, suicidal ideation, and coping strategies.
d. **Games:** This section is based on simple ice-breaking activities. Games maintain the same enthusiasm in entire therapeutic sessions.

Using self-management support in primary care can have a positive effect on the care and health outcomes of red collar employees with internet addiction after the pandemic.

9.5.6 STATISTICAL ANALYSIS

After data collection obtained data of pre-test and post-test of PMRT and SMT were analyzed by using SPSS.20 software. The obtained data were analyzed using Means, SD, and ANOVAs.

9.6 RESULTS

- Part A: Table 9.1 (Part A) reveals the scores of means and SDs for internet addiction across gender (males and female), conditions (pre and post) and therapy (autogenic relaxation training and SMT) among male and female red collar employees.

  Table 9.1 Part (A) shows that the means of Autogenic Relaxation Training for pre and post-test conditions were, respectively 19.43 and 11.7 while means of SMT for pre and post-test conditions were, respectively 18.45 and 10.22. This shows that the mean score of internet addiction after SMT was found to be less than Autogenic Relaxation Training.

  It is also cleared the inspection of Table 9.1 that the post-test mean of males and females were, respectively 10.55 and 11.37. This shows that the mean of females was greater than males.

TABLE 9.1 Part A: Mean and SD on Internet Addiction Among Red Collar Employees

Source	SS	df	MS	F	Sig.
Gender	313.290	1	313.290	14.410	0.000
Therapy	151.290	1	151.290	6.959	0.009
Gender × Therapy	3.240	1	3.240	0.149	0.700
Errors (*BS*)	4261.18	196	21.74	–	–
Conditions	6368.040	1	6368.040	1266.860	0.000
Conditions × Gender	90.250	1	90.250	17.954	0.000
Conditions × Therapy	6.250	1	6.250	1.243	0.266
Conditions × Gender × Therapy	10.240	1	10.240	2.037	0.155
Error (*WS*) (conditions)	985.220	196	5.027	–	–

- Part B: Table 9.2 (Part B) reveals the result of Analysis of Variance for Internet addiction across all the three variables under study. The main effect of Conditions, the main effect of gender and the main effect of therapy was found to be significant at the 01 level. The interaction effect of conditions (Pre and Post) and Gender (Male and Female) on internet addiction was found to be significant at 0.01 level while the interaction effect of Conditions (Pre and Post) and Therapy (ART and SMT) and the interaction effect of Gender (Male and Female) and Therapy (PMRT and SMT) was found to the insignificant. The combined interaction effect of Condition (Pre and Post-test), Gender (Male and Female) and Therapy (ART and SMT) was also found to be insignificant on Internet addiction.

TABLE 9.2 Part B: Summary of ANOVA on Internet Addiction among Red Collar Employees

	PMRT				SMT				Total			
	Pre-Test		Post-Test		Pre-Test		Post-Test		Pre-Test		Post-Test	
	Mean	SD	Mean	SD	Mean	SD	Mean	SD	Mean	SD	Mean	SD
Male	17.82	3.90	11.36	3.55	17.34	4.39	9.74	3.62	17.58	4.14	10.55	3.58
Female	21.04	3.92	12.04	2.60	19.56	4.39	10.70	2.27	20.3	4.15	11.37	2.43
Total	19.43	3.91	11.7	3.07	18.45	4.39	10.22	2.94	–	–	–	–

9.7 DISCUSSION

9.7.1 INTERNET ADDICTION IN RELATION TO GROUP THERAPY

The hypothesis of the present study was that:

> *"The respondents with ART and SMT will differ significantly from each other on Internet addiction."*

To test the above hypothesis Means, SDs, and F ratios were computed for internet addiction. The obtained Mean, SD, and F ratio for Depression are shown in Tables 9.1 (Part A) and 9.2 (Part B). As it is clear from the inspection of Table 9.2 (Part B) that the main effect of therapy was found to be significant at 0.01 levels. It is clear from the observation of Table 9.1 (Part A) that SMT was found to be more effective (mean being 14.33) than autogenic training (mean being 15.56) in reducing internet addiction among respondents. So, the hypothesis has been accepted.

As it is clear from the results that SMT was found to be more effective than Autogenic Relaxation Training even in reducing stress and internet addiction. The reason may be attributed to the fact that a key skill in self-management is self-regulation Self-Control alludes to people observing, controlling, and coordinating perspectives of their learning for themselves (Gupta & Srivastava, 2018). The leading utilize of self-management bolster is the collaborative interaction between the specialist and the client. Propelling, tuning in, and coaching are critical self-management bolster abilities that can make the specialist client interaction more grounded and in which all individuals of the care group can get to be learned. Through progressing preparing and hone, supporting clients in self-care will end up portion of day-to-day care. SMT is effective for building greater awareness and self-acceptance, changing unsatisfying patterns and behaviors, motivation, and developing the necessary skills to face life's challenges. This training also helps to live in the present moments and open true nature, which encompasses the inherent qualities of happiness and well-being (Cohen, 2006).

As it is well known that the SMT is a kind of experiential and motivational learning. These practices direct a person on the path of happiness, compassion, selflessness, and change negative thoughts into positive thinking which help a person to cope more effectively with depression, stress, and internet addiction. Thus, SMT is the process of realizing that every individual possesses unique qualities which differentiate him from others.

The SMT is found much effective because by this training, the changes are found in cognitive level also while in PMRT the changes of cognitive level are not found.

The results of Singh (2006), Nandi (1994); and Lingaswami (1992) have supported the results of the present study that an increase in the level of SMT significantly reduces internet addiction of Indian youth. The results are also supported by the study of Das et al. (2013), who found that Motivational Therapy significantly reduces the level of stress and internet addiction among adults. Duke University Medical Center (1998); and San Francisco General Hospital (1983) studied the effect of deep breathing along with motivational counseling on depressive and internet addict patients and found the significant positive effect of deep breathing and motivational counseling. Hannig (2016) wrote about motivational therapy that, it is a process like counseling, very close to the Catharsis, an emotional release.

9.7.2 INTERNET ADDICTION IN RELATION TO GENDER

The Hypothesis 2 of the present study was that:

> *"There will be a significant gender difference on Internet addiction among male and female red collar employees."*

To test the above hypothesis Mean, SD&F ratios were computed for Internet Addiction. The obtained Mean, SD&F ratios for internet addiction are shown in Tables 9.1 (Part A) and 9.2 (Part B). As it is clear from the inspection of Table 9.2 (Part B) that the main effect of gender was found to be significant at 0.01 levels. It is clear from the observation of Table 9.1 (Part A) that females had more internet addiction (Mean being 20.3 and 11.37 in pre and post-test conditions) than Males (Mean being 17.58 and 10.55 in pre and post-test conditions). The results support H2.

The reason may be attributed to the fact that females still have secondary position in society. Situations become acute for females than males. Household responsibilities and workplace environment are strongly correlated to women's work-life conflict they cannot manage everything thing, and they are unable to express their emotions which makes them feel more anxious and depressed as compared to males. Since no one seems to understand them, they tend to internalize their emotions. Although the trend is changing rapidly but the gender discrimination has so deeply rooted that it operates in each and every sphere of life. Gender stereotyping and bias is rampant in almost all walks of life be it in primary Rhymes, or Color Preferences or at professional front. In family gender differences operates even in the division of household works. The boys will either be not assigned any work or if assigned they will be expected to do such chores as buying necessary items from market, etc., but girls are given typical works which are conventionally meant for females such as dusting, cooking, etc. So it operates at every level therefore females are not generally encouraged to talk about themselves and to express how they feel so even when given the opportunity they often have trouble verbalizing their emotions. The gender difference in depression and internet addiction is one of the most robust findings in psychiatric epidemiology and it's very much associated with internet addiction (Norman et al., 1989). These results are supported by Survey of Voice of Azamgarh Community Radio (2018) which was done on working females of government sector. In this survey, the main reason for working women Internet addiction and depression was contributed to the fact that they were treated as secondary citizens in the society, they lack their own identity, thoughts, desires, and freedom. These females don't have many opportunities to share their emotions. They cannot even share or express their physical and mental health problems, so they were engaged on the internet to express their feeling on various social media platforms.

The present finding is also supported by the study of Piccinelli & Homen (2018) who reported that depression contributes most significantly to the global burden of disease and it most frequently encountered female's mental health.

9.7.3 INTERNET ADDICTION IN RELATION TO CONDITIONS

As it is evident from the inspection of Tables 1.1 Part (B) that the main effect of conditions (pre and post) was found to be significant on Internet addiction. So, hypothesis related to the conditions was accepted. These results suggest that intervention is significantly affecting the Internet addiction, among male and female red collar employees because before the intervention the respondents were having more internet addiction, but after intervention it reduced a lot. The results of the present study are supported by the study of Verma & Lal (2018) who have suggested that exercise and increased physical, motivational, and life skill's activity improves function, quality of life, strength, endurance, and reduced excessive use of internet.

9.7.4 INTERACTION EFFECT

i. **Internet Addiction in Relation to Gender and Group Therapy:** No Significant Interaction between Gender and Therapy was found on Internet addiction.
ii. **Internet Addiction in Relation to Condition and Gender:** Depression was reported less among males than females in post-conditions of both the therapies (ART & SMT).
iii. **Internet Addiction in Relation to Condition and Group Therapy:** The interaction effect of Condition and Therapy was not found to be significant on Internet addiction.
iv. **Internet Addiction in Relation to Condition, Gender, and Group Therapy:** The interaction effect of Conditions, Gender, and Therapy was not found to be significant on Internet addiction.

9.7.5 CONTRIBUTION

The very important contribution of the study is that it not only tries to assess the effectiveness of Group Therapies, but it also compared the effectiveness of two very important and useful Group Therapies, i.e., ART, and SMT in dealing with Internet addiction.

The main contributions of the present study are as follows:

- The results of the present study may contribute significantly to the existing body of the literature since very little work has been done till date together on these variables and very few attempts have been made to explore the effectiveness of different intervention trainings on managing and coping internet addiction especially faced by red collar employees. Results of the present study can guide counselors and policy makers to help the related group accordingly.
- Since the study is done on red collar employees it become even more enlightening for the future researches because it is the need of today's time to understand the life style and psychology (i.e., how employees perceive and feel about certain life situation) of the employees since this is the sample of the population which may be considered as vulnerable group.
- The next very important contribution of the study is that it not only tried to see the effectiveness of Group Therapies but it also compared the effectiveness of two very important and useful Group Therapies, i.e., PMRT and SMT in dealing with Internet addiction. The findings have suggested that SMT was found to be the more effective intervention training as compared to PMRT. These results are very enlightening since most of the past literature has suggested PMRT as a very effective technique in reducing stress, negative thoughts and internet addiction, enhancing Coping Strategies, but the findings of the present study have suggested that the combination of different therapies can even lead to better intervention. The reason may be attributed to the versatility of SMT. It comprises many techniques which need active participation by the respondents generating more interest and motivation among them so they don't need any external push to deal with their problems.
- The study also shed light on the fact that we not only deal with internet addiction by strengthening positive coping strategies but also deal with internet addiction effectively by weakening on negative coping strategies. So, there is a need in the present time, to promote positivity and discourage negativity among people by proper counseling so that people will be able to see the same world in a better manner from a different and healthy perspective. SMT was found to fulfill this purpose effectively.

The present study has tried to mitigate the effect of internet addiction, build a healthy and balanced life among red collar employees through Group Therapies and tried to make them more productive employees specifically and contributing citizens of society generally.

9.8 SUGGESTIONS FOR FUTURE RESEARCH

In the light of above said contribution and limitation of study there are some suggestions for future researchers:

- For conducting this study, the investigator selected only a few psychological variables, but when the results were analyzed, it was felt that some other variables should also be taken. These may be gaming addiction, harassment, life satisfaction, aggression, etc.
- The mere provision of information about the negative consequences of risk behavior is ineffective, and it needs to be complemented by interactive interventions aimed at changing attitudes and the development of selected skills for life.
- Random sampling should be used to make the study more reliable.
- Regular counseling sessions by trained counselors should be organized or provision of the mental health clinic in the organization for employees should be made. Through this study it is highly recommended to the responsible authorities.
- This study should be done with other employees like white/blue/pink/gray/gold/black collar employees. It will help to explore the tendencies of different employees in various environment, thoughts, and socio-cultural atmosphere.
- More helpful methods can be utilized for superior comes about. The successful preventive intercessions ought to be complex and centered on Web enslavement and other shapes of chance behavior and ought to be a combination of mediations focusing on vulnerable individuals with an Online addiction, their parents and other adored ones, and the community, school, or work environment.

9.9 CONCLUSION

Despite these limitations, the findings of the study have suggested that SMT was found to be the more effective intervention training as compared to ART. The outcomes of the present study have revealed that the combination of multiple therapies can result in an even better intervention, which is extremely illuminating given that the majority of previous work has suggested ART as a very successful strategy in reducing stress and internet addiction. The adaptability of SMT may be to account for this. It includes a variety of strategies that demand active participation from the respondents, increasing their interest and drive so they don't require any outside encouragement to solve their problems. As a communicable illness, and presently a worldwide public health emergency (PHE), COVID-19 places a one of a kind challenge on our wellbeing and social care workforce that will disturb not fair their regular work environment obligations but moreover their social setting and combined treatments may be the reply.

9.10 SUGGESTIVE MEASURES

- Making an movement plan for each day and week (e.g., arranging in development when one is progressing to work/study, lock in in social exercises, perform recreation exercises, conduct physical work out) and advancing an everyday schedule at domestic in self-isolation, isolate, or lockdown may be exceptionally supportive at a time when every day structure is misplaced or reduced.
- Sleeping routinely and sufficient, eating frequently and soundly, drinking adequate liquids, and attending to individual cleanliness are basic not as it were to preserve good physical wellbeing but moreover for upgrading mental well-being.
- Engaging in physical movement frequently isn't as it were vital to keep a sound body but moreover contributes to boosting mood by diminishing levels of stress hormones, invigorating the generation of endorphins (i.e., common chemicals within the brain that calm torment and improve disposition) and having a useful impact on resistant work. Learning and using relaxation and other stress-reduction techniques (e.g., reading, writing, listening to music, meditation, autogenic training, and mindfulness exercises) can be helpful in keeping bodies and minds healthy and to be aware of one's emotions. When dealing with difficulties, openly communicating about emotions with a close relative or friend, asking for help and feeling social support can effectively help to reduce stress and anxiety.

- Getting a charge out of social exercises and keeping up connections are too pivotal. Family individuals ought to orchestrate to spend quality time with each other and have "family time" periods on a normal premise. Family time may incorporate important discussions, playing social recreations or sports, eating suppers and doing family chores together.
- Many studies have recommended that intercessions with family individuals or other relatives like "Community Fortification and Family Training" can be valuable in improving the inspiration of an someone who is addicted to cut back on internet utilize, in spite of the fact that the reviewers comment that control thinks about with relatives don't exist to date.
- Reality therapy (RT) is assumed to empower people to select to move forward their lives by committing to change their behavior. It incorporates sessions to appear clients that enslavement may be a choice and to donate them preparing in time management along-side SMT.
- For people who live at the side their family or others, it may be valuable to discover ways of being alone or having some self-time regularly. It may offer assistance to arrange spaces within the house for person and common utilize, as well as to set up and regard boundaries such as doors being open or closed. This may offer assistance diminish frustration and clashes that will emerge from being kept within the same put for long periods.
- As we alter to better approaches of living and working, red collar employees are likely to proceed to confront challenges ahead. Our inquire about affirms that the mental effect of internet addiction after COVID-19 on red collar workers is impressive, but Self-Management interventions can be conveyed after widespread conditions to moderate deteriorations in mental well-being and bolster mental health.

KEYWORDS

- **autogenic relaxation training**
- **COVID-19**
- **Internet addiction**
- **Internet addiction test**
- **public health emergency**
- **reality therapy**
- **red collar employees**
- **self-management training**

REFERENCES

Abir, T., (2021). *Internet Use Impact on Physical Health During COVID-19 Lockdown in Bangladesh: A Web-Based Cross-Sectional Study.* NCBI. Retrieved from: https://www.ncbi.nlm.nih.gov/pmc/articles/PMC8535707/ (accessed on 7 February 2023).

Bayih, et al., (2020). Impact of social media on administrative staff employees in case of Woldia university. *International Journal of Innovative Science and Research Technology, 5*(6), 614–619. Retrieved from: https://ijisrt.com/assets/upload/files/IJISRT20JUN423.pdf (accessed on 7 February 2023).

Beck, A. T., (1961). *Depression Inventory, Depression: Clinical, Experimental and Theoretical Aspects.* New York: Harper & Row.

Beck, et al., (1979). Assessment of suicidal intention: The scale for suicidal ideation. *Journal of Consulting and Clinical Psychology, 47*(2), 343–352.

Bracke, (2000). *Mental Health Among Drug Abusers* (pp. 11–33). Brunner/ Mazel Int, New York.

Cohen, A., (2009). *N*eglect and perceived stigmatization impact psychological distress of orphans in Tanzania. *European Journal of Psycho Traumatology*. Retrieved from: http://doi.org/10.3402/ejpt.v6.28617.

Cuncic, A. (2020). Autogenic Training for Reducing Anxiety. Retrieved from https://www.verywellmind.com/how-to-practice-autogenic-training-for-relaxation-3024387.

Fludd, V., (2014). *Workplace Productivity Drains In: T+D, 68*(9).

Grodzinsky, et al., (2005). Internet and productivity: Ethical perspectives on workplace behavior. *Journal of Information Communication and Ethics in Society, 3*, 249–256.

Gupta, N., & Srivastava, N., (2018). Suicidal ideation among orphans and its management through group therapy: A comparative study. *Research Journal of Social Science & Management RJSSM, 7*(11).

Hannig, G., (2010). A randomized clinical trial of alprazolam versus progressive muscle relaxation in cancer patients with anxiety and depressive symptoms. *Journal of Clinical Oncology, 9*(6), 1004–1011.

Harry, E., (2017). Effect of CBT on psycho-physical functions. *Journal of Indonesia Psycho, 23*(1), 37–42.

Henary, et al (2018). Pattern of chronic diseases amongst adolescent obese children in developing countries. *Curr. Sci., 88*, 1052–1056.

Hermenan, et al., (2018). Cross-culture aspects of self-management in Asian girls. *Nutrition & Food Science, 96*(4), 19–22. Retrieved from: https://www.theinternationaljournal.org/ojs/index.php?journal=article&-view&path%5B%5D=6802&path%5B%5D=pdf (accessed on 7 February 2023).

Jorm, et al., (2019). Eating habits and factors affecting food choice of adolescents living in rural areas. *Hormones, 12*(2), 246–253.

Kimberl, et al., (2010). Versatility and addiction in gaming: The number of video-game genres played is associated with pathological gaming in male adolescents. *Cyberpsychology, Behavior & Social Networking, 18*(2), 129–132. doi: 10.1089/cyber.2014.0342.

Lingaswami, (1992). Drug abuse, dependence, and addiction in chronic pain patients. Department of Psychiatry, University of Miami, School of Medicine, Florida. *The Clinical Journal of Pain, 8*(2), 77–85.

Lucy, K., et al., (2015). A study of psychophysical health in relation to physical conditions of respondents. *Indian Journal of Health & Wellbeing, 2*(1), 115–118.

Minkoff, T., et al., (1973). Does mindfulness medication improve anxiety and mood symptoms among alcoholic. *Canadian Journal of Psychology & Behavioral Science, 52*(4), 260–266.

Mukhtar, et al., (2021). The effects of social media use on preventive behaviors during infectious disease outbreaks: The mediating role of self-relevant emotions and public risk perception. *Health Commun., 36*, 972–981. doi: 10.1080/10410236.2020.1724639.

Piccinelli, et al., (2018). Internet addiction, usage, gratifications, and pleasure experience – The Taiwan college students' case. *Computers & Education, 35*(1), 65–80. doi: 10.1016/S0360-1315(00)00019-1.

Pitichat, T., (2013). Smartphones in the workplace: Changing organizational behavior, transforming the future. In: *A Journal of Trans-disciplinary Writing and Research from Claremont Graduate University* (Vol. 3, No. 1, pp. 1–11).

Sharma, N., (2018). Nutritional awareness among the parents of primary school going children. *Saudi J. Humanities Soc. Sci., 2*(8), 708–725.

Singh, L., et al., (2016). Influence of grandparents on depressive behaviors of young children in Chinese three-generation families. *Science Direct, 48*(3), 377–383.

Srivastava, N., et al., (2021). Mental health in relation to internet addiction among Indian University Students. *Vidyabharti International Interdisciplinary Research Journal (VIIRJ), 13*(1), 445–451. Retrieved from: http://www.viirj.org/vol13issue1/65.pdf (accessed on 7 February 2023).

Tizazu, et al., (2020). A cognitive-behavioral model of pathological internet use. *Computers in Human Behavior, 17*(2), 187–195. doi: 10.1016/S0747-5632(00)00041-8.

WHO, (2019). *Young People's Health: A Challenge for Society, WHO Tech. Report, Geneva*. Retrieved from: http://www.who.int/countries/ind/en/ (accessed on 7 February 2023).

Young, et al., (2019). *Internet Addiction Test*. Retrieved from: https://www.iitk.ac.in/counsel/resources/IATManual.pdf (accessed on 7 February 2023).

CHAPTER 10

Paradigm Shift in Consumer Shopping Behavior During the COVID-19 Pandemic

PRIYA VIJ,[1] KASHISH KAKRA,[2] and ATTIN KUMAR[3]

[1]*Professor, Economics, Chandigarh University, Punjab, India*

[2]*IND Analyst II-HCS, Aon Consulting Pvt. Ltd., Gurgaon, India*

[3]*Cluster Manager, Reliance Trends, Gurgaon, India*

ABSTRACT

The world during the COVID-19 pandemic has witnessed massive changes in consumers' choices, and in the way, businesses are responding to sustain their margins. The use of the Internet has grown significantly, particularly as means of communication, entertainment, and commercial exchanges. Covid-19 has accelerated the pace of digital transformation prioritizing digital infrastructure as the top policy agenda of the government. Today's consumers are getting adapted to the online mutation and are demanding personalized offerings. However, this tremendous growth has been accompanied by rising apprehensions among business firms to sustain their bottom line. In response to the prominent structural changes in consumer behavior, business firms are adapting an omni-channel retail strategy and are improving their offerings and policies. This chapter throws light on the changing trends in consumer shopping behavior during the COVID-19 pandemic, the impact on businesses with special reference to the FMCG sector, the increasing role of e-grocers, online foods delivery

Reshaping the Business World Post-COVID-19: Management Strategies for Sustainable Behavior Change. Arvind K. Birdie & Ruchi Joshi, (Eds.)

providers and impact on taxi services in terms of the challenges they faced and their response to the change and finally addressing consumer privacy and security concerns in online shopping/booking orders. The chapter discusses the policy implication with respect to consumer and businesses adaptation to the New Normal.

10.1 INTRODUCTION

The World Health Organization has proclaimed COVID-19 to be a pandemic. The pandemic's rapid spread has thrown the global economy to a halt. It has created a lot of uncertainty and has had an influence on people's lives. Millions of firms are on the verge of going out of business. Almost half of the world's 3.3 billion workers are facing job losses. The majority of workers in the informal economy lack social security, access to decent health care, and productive assets, making them particularly vulnerable. This pandemic has had a major impact on global supply chains.

The international community's first response to COVID-19 was to close borders followed by restrictions on domestic mobility. Social distancing and Quarantine norms became the new order for most of the countries of the world with varying restrictions. Some countries like China, Iran, Italy, Spain, France, Belgium, and Portugal implemented strict domestic measures as compared to the United States, Canada, Brazil, Russia, Netherlands, Austria, Switzerland, and Israel which did not impose strict domestic travel restrictions. India too took measures to contain the spread of the pandemic including imposing a lockdown from 25th April 2020. There was a full lockdown at first, followed by a partial lockdown, which meant that most people were forced to stay at home and were only allowed to leave the house for needed products from grocery stores or pharmacies. Many businesses closed, and many individuals lost their employment as a result. The second wave spread more quickly across countries than the first, and it was significantly more severe, affecting more than 200 countries. And now even in the third year of the COVID-19 pandemic, with the rise of another variant 'Omicron' world economies are again experiencing the fear and turmoil.

Since 1870, the global economy has gone through 14 global recessions. According to current projections, the global recession of COVID-19 will be the fourth deepest and most severe since the Second World War. A historically high proportion of countries is thought to be experiencing per capita production reductions. In the third quarter of 2019–2020, India's real GDP fell to its lowest level in over six years, and the emergence of COVID-19 brought forth new concerns. Private consumption, investment, and external trade, the three key drivers to GDP, will be impacted. The primary, secondary, and tertiary sectors will witness the impact of the pandemic.

1. **Agriculture:** In Agriculture due to a global market downturn for hotels and restaurants, agricultural product prices are expected to fall by 20% in 2020. The supply chains have failed to function correctly, and vast amounts of food have begun to go to waste, farmers will suffer enormous losses as a result of this. In light of the pandemic, which is already dampening oil demand, the oil-price war is expected to have significant consequences for the global economy.
2. **Manufacturing Industry:** In the manufacturing industry, due to the insecurity of supply chains and self-isolation policies, import issues and personnel shortages have emerged as major challenges for businesses.
3. **Education Sector:** In the education sector, according to UNESCO, the closure of educational institutions has affected up to 900 million students. Because schools are no longer able to provide free school meals, COVID-19 has severely hit social mobility, leading to social isolation and higher school dropout rates for children from low-income households.
4. **Healthcare Sector:** In the healthcare sector, the pandemic has created unparalleled problems for the healthcare systems worldwide. One of the most serious flaws in healthcare systems around the world is the high risk to healthcare personnel.
5. **Hospitality, Tourism, and Aviation:** For hospitality, tourism, and aviation, the tourism industry is currently one of the hardest hits by the COVID-19 outbreak, which is affecting both supply and demand for travel. The World Travel and Tourism Council has warned that COVID-19 might jeopardize 50 million jobs in the global travel and tourism industry.

Despite the fact that vaccines are available in 2021, current vaccine arrangements have raised hopes for a positive economic impact later this year. However, new outbreaks as well as virus variants pose challenges. This pandemic will forever change the way businesses function and consumers behave. During the pandemic, many retailers have embraced digitalization and are making themselves available online and offline meeting the need of the time. According to recent research by the retail tech company SnapBizz, over 1 million unorganized shops in India have digitized their operations and are now able to fulfill client orders by delivering doorstep delivery. Many companies have developed customized apps to make internet buying easier for customers. Consumers are more resilient in shopping online for all groceries and household needs during the pandemic, according to one survey. Consumers and businesses have been able to withstand the epidemic, with due credit to widespread social networking. As per one of the reports, OECD countries are making increasing use of ICT tools to take advantage of the transition to online behavior during the pandemic.

The pandemic has made businesses shift from a growth mindset to a profit mindset: Due to the pandemic retailers are re-thinking on their business models and shifting away from a "growth at all costs" idea towards being able to survive and maintain their profits. The goal is to ensure that businesses are still alive. Businesses are adopting to new workplace norms, such as remote work, increased video conferencing, leaner workforces, and more automated processes, which will most likely persist long after the pandemic has passed. Businesses are now planning as to how to incorporate these new work styles and habits as a part of normal routine.

The demand for new technology and the need for making quick decisions during the pandemic led to technical advancements in several fields. Consider how rapidly people acclimated to virtual open houses and how senior citizens adopted to internet banking and shopping. Consumers are experimenting with new ways of doing things that would have taken a long time to adapt in the past. In light of these behavioral changes, the present chapter focuses on the following three objectives:

- Analyzing the consumer shopping behavior during COVID-19 pandemic;
- Understanding the impact on businesses with special reference to the FMCG sector, e-grocers, online foods delivery providers and taxi services across the globe; and
- Understanding the response of businesses to the challenges posed by the COVID-19 pandemic.

10.2 CONSUMER SHOPPING BEHAVIOR DURING THE COVID-19 PANDEMIC

The COVID-19 outbreak and subsequent lockdown had a tremendous influence on the global economy, particularly on consumption, which accounts for a major part of the Gross Domestic Product of any country. The economic growth of the world being less than 2.5%, some countries experienced recession as a result of the pandemic. Buying things from traditional local retailers has been continuing since long especially in reference to developing countries. Many buyers prefer to shop offline so they may wish to try the product before they purchase. Behavioral shifts in buying behavior are seen throughout the world as a result of the pandemic. Consumers are becoming more digital and prefer to shop online rather than in real places. "A growing share of internet users are buying day to day essential products, including as food and beverages, cosmetics, and pharmaceuticals online," according to a UNCTAD and NetComm Suisse eCommerce Association research.

Huge overnight changes seemed visible during the pandemic, there has been a turnaround in the way What, When, and How consumers shop:

i. **What:** The major shift of consumer shopping during the pandemic was on food and beverages, gifts, apparels, toys, and home and garden, household cleaners and soap, sanitizers, disinfectants. What people are not buying: Luggage, electronics, Gym bags, cosmetics, and sun care products.
ii. **When:** Consumers are buying in bulk to stock up things in the light of the lockdown uncertainty.
iii. **How:** Shopping trends have been shifting from offline to online for a long time, but the COVID-19 pandemic gave an instant boost to online business and shifted the majority of consumers to it.

People are more likely to buy online now because they feel more at ease. Some of the prominent behavioral trends in consumption which have been witnessed are:

1. **Bulk Buying/Panic Buying:** It is defined by Paul Marsden, a consumer psychologist, as "playing to our three fundamental psychology needs of autonomy (need to feel in control of your actions), relatedness (need to feel that you are doing something to benefit your family), and competence (need to feel like smart

shoppers making the right choice)." In anticipation of the looming uncertainties, consumers have started storing up on necessities.

2. **Scheduled Shopping:** These where grocery shopping became a planned affair for most consumers during the pandemic.
3. **Digital:** It is now more significant than ever before. Shift to online shopping, online entertainment and online food delivery.
4. **Lifestyle Changes:** These were observed in the form of healthy food and Yoga routine. Customers have been spending more on their health and wellness, to stay healthy and face the coronavirus. The purchase of vitamins and supplements increased.
5. **Demand of Essentials:** There has been a drop in overall consumption, where major purchases constituted essential products. Customers have shifted to value buying. More demand for fruits and vegetables and dairy products than junk food.
6. **At-Home Entertainment:** This was the only option available which gave a boost to the OTT (*over-the-top*) media services offered over the internet to suit the requirements of individual customers.
7. **Cut in Travel:** Out of home entertainment, purchase of fashion and luxury goods, durables, and appliances, etc.
8. **Mixed Behavior Patterns:** These were observed with respect to alcohol and online food delivery.
9. **Brand Loyalty:** It has been shaken.
10. **Home Grooming Solutions:** These like hair color, epilators, etc., increased.
11. **Demand for Repair Services:** It has shot up mostly due to consumers' desire to extend the life of their existing white goods rather than purchase new ones.
12. **Increased Usage of Communication Platforms:** According to one survey, WhatsApp, Instagram, and Facebook Messenger, all owned by Meta, were the most widely used communication tools. The use of video conferencing tools in the workplace has benefited Zoom and Microsoft Teams the most.
13. **Still Local:** Despite the above trends, customers are still willing to buy from local shops closer to their homes, and this has given the opportunity to organized retail sector to open stores in Tier 2 and 3 cities.
14. **Increased Investment:** It has been seen in Jewelries: E-gold and E-diamonds.

In light of the behavioral shift in consumer preferences as seen during the COVID-19 pandemic, in a contemporary world, it becomes vital for businesses to meet customer expectations with respect to delivery, along with ensuring quality and value for money. And leveraging the technology to its advantage will enable retailers to reach their customers much faster and in an easier and economic way, thus adapting to the transformation.

10.3 IMPACT ON BUSINESSES

The following are some of the issues that businesses experienced during the pandemic:

1. **Shopping Restrictions:** Social distancing, non-availability of changing rooms, intense cleaning procedures, quarantining goods, and the restrictions on multiple customers to try the same thing at the same time are some of the challenges which retailers faced during the pandemic.
2. **Changes in Consumer Basket:** Originally, online shopping was more popular for fashion and makeup brands, but the consumer basket has shifted from fashion to necessities such as food, medicines, and heath related products.
3. **Supply Chain Issues:** Many small enterprises have been compelled to limit operations or temporarily close plants as a result of the corona virus epidemic. For example, as part of a global supply chain shift, the epidemic is placing pressure on businesses to shift their operations away from China and look for opportunities in other emerging economies.
4. **Unpredictable Economic Outlook:** With multiple variants of COVID emerging, predicting the flow of economic trends is extremely difficult.
5. **Offering Discounts/Sales:** Businesses are offering huge and appealing discounts in order to keep their customers and keep their businesses moving.
6. **Marketing Through Social Media:** People's screen time has increased in lockdown as they have become more active on social media, making it an excellent platform for advertising products. Businesses are trying to understand the behavioral changes to target their customer through social media.

7. **Safety Concerns:** Consumers will be more concerned about safety and hygiene and will increased restrictions on their travel to malls and crowded places will lessen. Thus it will result in decreased store visits and increased online shopping.
8. **Liquidity:** Consumers priority will be health, maintaining hygiene, thus major wallet share will be taken by products that maintain well-being or health, for infection containment (e.g., face masks) and higher purchases of shelf-safe products.
9. **Fashion:** Because it relies so heavily on discretionary spending, the fashion industry is most vulnerable to disruption from the corona virus outbreak. Consumers will be more concerned with meeting their basic needs for healthcare, hygiene, and other essential requirements; therefore fashion will take a back seat. Though the pandemic's duration and final severity levels are still unknown, it is clear that the fashion industry is only beginning to struggle.
10. **Limited Spend:** The crisis is influencing our everyday lives, causing anxiety and uncertainty in practically everyone's minds. Consumers will choose to buy only what they require rather than what they desire. The discounted wave may have differing degrees of influence on different market categories.
11. **Shift to E-commerce:** Only businesses or retailers who are adapting to the online model are more able to survive the pandemic. Therefore, it necessitates businesses and retailers to transform too in light of the behavioral changes in consumption.

The subsections will explain the impact of COVID-19 pandemic with special reference to the FMCG sector, e-grocers, online foods delivery providers, and Ola and Uber services across the globe.

10.3.1 IMPACT ON THE FMCG SECTOR

The outbreak of the COVID-19 pandemic has triggered a global crisis in the fast-moving consumer goods (FMCG) industry. The FMCG industry has been severely hit by social distancing norms. Issues with respect to supply chain and logistics, unavailability of labor, operations being limited to manufacturing of only essential commodities, and so on are just a few of the challenges that most FMCG companies faced across the world.

Stay-at-home consumption with an emphasis on health and wellbeing, as well as sustainability, dominated the strategic agenda in the FMCG industry.

During the pandemic lockdown, ecommerce saw a boom in sales of FMCG, as people were afraid to leave their homes for fear of exposure to the virus. There was a significant rise in demand for FMCG goods, sanitizers, and other essential items even before the lockdown was declared by the government. According to FMCG businesses, demand increased suddenly between 15% and 45% for various categories of goods. Basic essentials such as Rice, atta, pulses, edible oil, sugar, tea, biscuits, noodles, frozen food, soaps, hand washes, and floor cleaners were in high demand. Even the small and medium retail shopkeepers are going online by creating their own websites and also by joining big retailers such as Amazon, Flipkart, Grofers, and other e-commerce platforms to enhance sales and meet consumer demands.

Because of the scarcity of fresh fruits and vegetables, companies that make processed foods and non-perishables, for example, have witnessed a significant rise in sales. While most FMCG sub-sectors benefited from favorable consumption trends during the early stages of lockdowns, there were several notable exceptions.

10.3.2 IMPACT ON E-GROCERS

As per an industry trade publication dated March 2020, approximately 40 million (31%) American consumers switched to online grocery purchase. This is more than double the quantity recorded in the August 2019 poll. According to the poll, 26% of users switched to online services for the first time. Monthly order rates grew 19%, while total monthly order volume increased 193% among users. Most retail categories will use eCommerce as their principal channel of commerce until the ban on shelter-in-place orders is repealed, with buy online, pick up in store (BOPIS) and other techniques supplementing it. Consumers will eventually return to physical stores, but they will still hold a sizable market share.

E-grocers including Amazon, Flipkart, BigBasket, NaturesBasket, Grofers, Walmart, and Meijer have seen a 100% increase in sales, with staples, biscuits, frozen soaps, and handwashes seeing even greater growth. The FMCG sector has been working to speed up delivery of goods ensuring no shortages. Panic buying was observed during the pandemic, with everyone buying in large amounts in order to store the goods for longer period of time. As a result, it became increasingly challenging for FMCG companies to meet consumer demand, as orders came from both local retail stores and online grocers. According to Mr. RS Sodhi (Amuls' Managing Director), some FMCG businesses such as ITC Ltd, Parle Products, Amul, and Godrej were running out of stock, and most of the goods didn't last even a week. Because of the high demand for FMCG goods online, shortage of delivery personnel was seen. Online grocers like BigBasket, Grofers, and NaturesBasket all were short of delivery boys/ girls, and as a result, these companies were given 2 weeks or sometimes 3 weeks delivery dates. More and more influx of new online shoppers was People who had never used an online shopping platform before were also signing up and placing orders.

The pandemic also saw Amazon and Flipkart, the biggest online retailers in India, as facing stiff competition with BigBasket of Tata Group and Grofers that are fulfilling a large amount of grocery demands in the country now. New players, Swiggy and Zomato have also joined the race recently to make things even tougher for the two dominating e-commerce companies. JioMart, an Indian e-commerce company also ventured during the pandemic in December, 2019 to cater to the grocery needs of Indians. JioMart is a joint venture between Reliance Retail (India's biggest retail chain) and Jio Platforms and is now catering to a market share of 2%. Also, Future Groups' BigBazaar started catering grocery online orders from JioMart portal thereby reaching to a wider market with a TAT (turnaround time) of maximum 24 Hours. Grofers now Blinkit is assuring its customers delivery within 10 minutes at customers' doorsteps. Hence, the COVID-19 pandemic has led massive structural transformation in the consumer shopping behavior.

10.3.3 IMPACT ON ONLINE FOOD ORDERING

The lockdown restrictions immediately changed the way people looked at food purchases. Firstly, the closure/restrictions at restaurants and eating places and secondly the change in food habits during the pandemic brought a change in preference towards online food ordering. For example, people preferred eating healthy three meals a day, and their frequency of shopping trips also lessened due to markets/malls being closed. The following changes were observed in the food category:

- All restaurants and bars were closed, with the exception of take-away and delivery services.
- The demand for online grocery shopping increased dramatically, causing the existing infrastructure to strain (evidenced by websites that crashed or operated extremely slowly and by long turnarounds in delivery time options from between 1 and 2 days to 10 days or more).

Online food delivery saw a massive rise with Zomato and Swiggy taking the lead and competing for market share. Zomato and Swiggy faced several hurdles across cities while trying to deliver food. Online food delivery orders dropped 70% as customers stepped back to place orders and top restaurants being shut amid a lockdown. Zomato understood the need of delivery of grocery, and it started online delivery of groceries in 80+ cities. It also partnered with Grofers for delivery.

Some steps to revive business during the COVID-19 pandemic include, subscription plan of Zomato Gold was extended. The Zomato Gold Support Fund helped the restaurant personal where annual subscription plan money was provided to them. Takeaways seemed an attractive option for customers, with a contact-less collection of food and with more discounts which led to reduced burden on customers and restaurants. Online delivery portals donated money through Rider Relief Fund to the delivery personnel since their earnings dropped with decrease in online food delivery. Also, the delivery partner insurance plan was launched aiming to help the delivery personnel in case they were infected by COVID-19, by providing them with a cover for hospitalization charges, and lost earnings.

10.3.4 IMPACT ON TAXI SERVICES

Taxi services as essential services have severely been hit by the COVID-19 pandemic. In addition to factors such as demographic, behavioral, socio-economic, and cultural, several research points out that population movement and mass transportation are directly linked to the transmission of infectious illnesses. With public transportation posing a greater danger of illness spread, the demand for taxi services, will majorly be impacted. More crucially, the nature of their employment, which may include transporting clients who may be ill or even being a healthcare worker, puts taxi/cab drivers at a higher risk of becoming infected. Also taxi drivers being vulnerable to infection also cannot afford to not work for a livelihood. They are put under a lot of financial and mental strain as a result of this. As a result, there is a global push to enforce necessary safety measures and checks in public transportation and taxis/cabs so that people can feel comfortable while traveling in taxis. The transportation sector has been one of the primary service sectors being affected by the pandemic, with the two major service players Ola and Uber reporting revenue losses.

Low demand and increased operating costs are causing problems for Ola and Uber cab drivers. There has been a drop in the revenues, and drivers are paying extra for sanitizers and disinfectants. There has been a drop in the revenues, and drivers are paying extra for sanitizers and disinfectants. Taxi service providers are providing personal protective equipment kit (which includes sanitizer, disinfectant and 10 marks) to their drivers or are reimbursing the amount, if purchased by the drivers. Ola has rolled out a 10-step approach for consumers and drivers to abide by to ensure COVID appropriate behavior.

10.4 THE RESPONSE OF BUSINESSES TO THE CHALLENGES POSED BY THE COVID-19 PANDEMIC

As discussed in the previous section, the impact of the pandemic has been perceived worldwide. Businesses must implement new technologies and strategies across the value chain to safeguard their business models in order to deal with new constraints, mitigate the pandemic's negative impact, and adapt to economic and consumer shifts. The following are some of the strategies:

- The transition from bricks to clicks (offline to online commerce model/omni-channel paradigm). To give customers a variety of ways to complete orders, including online shopping and in-store pickup.
- Understanding customer shopping behavior both online and offline.
- Digitization needs to be made more inclusive across sectors.

- Contactless delivery, where delivery personal drop delivery at a particular location, or in front of a door started.
- Retailers should aim at the provision of personalized offerings for their customers.
- Time to rewire the fashion systems which are no longer working.
- Focus on crisis management and contingency planning.
- Shift towards reimagining our industry altogether.
- Online companies to completely disable cash on delivery and encouraged digital payment to reduce contact.
- Strong leadership: Now is a time for leaders to work with the "new normal."
- Leaders need to be:
- Calm, confident, positive, and devoted to work;
- Empathetic: Show people we care and understand their situation. Know where our people are in the grief curve;
- Resilient and protect their employees;
- Setup a cross-functional COVID-19 response team;
- Making decision wisely with all thinking of uncertainties.

10.5 KEY INITIATIVES AND NEW MEASURES OF SAFETY: ADAPTING TO THE NEW NORMAL

Key initiatives by businesses during the COVID-19 pandemic include:

1. **Video Marketing:** Capturing hygiene and new measures of safety adopted inside the store to build up customer confidence. Intend of creating video and sharing it with all customers to establish the "safest retail store to shop in the country."
2. **COVID-19: Certified Sales Associates/Safety Champions:** Staffs training with respect to safety measures and staffs who will pass the exam, will be certified, and get a badge and a certificate.
3. **We Care for You:** Couriering mask to all registered customers to convey to customers that "your safety is our first priority."
4. **Changes in Shopping Process:** The shopping process must be revised with strict process change, and all processes must have only one thing common "security of individuals: customer and staff." Measures must be taken like:

i. Social distancing measures;
ii. Sanitization measures and other safety measures.

It is going to be an ongoing process and a lot more initiatives will be required from time to time to create demand and revive consumer confidence. All above said measures are required to be adopted at all touch points starting from security guard, baggage counters, trial rooms, CSA (customer service associate) interaction, billing, and exchange. These measures give confidence to the customers that place they are shopping are really safe to shop.

In today's tumultuous environment, some businesses have improved their ability to plan for and mitigate risk in the event of a crisis. They are become more resilient in their design and offerings. Unexpected situations like this are likely to emerge more often, requiring leaders to be more adaptable, transparent, and forward-thinking and be sustainable. In the ever-changing retail environment, the one that embraces change will always be one step ahead of the competition. The entire change management revolves around 5A's:

1. **Aspire:** When a change occurs, having a vision is essential. Everyone can see everything because of the vision. Your team members will understand why they are working on the change endeavor if you have a clear vision. Change is only achievable when you have a solid strategy in place and create a sense of urgency to meet your goals.
2. **Assess:** One need to identify the area of improvement with change in mind set of the staff for achieving the Vision.
3. **Architect:** This will include how do you change an individual in system and process, building skill and creating role model may be by leading example from front.
4. **Act:** Then comes acting, the same as creating energy among the team.
5. **Advantage:** This includes creating stories, creating urgency in the surrounding, building skills of the staff and development of an individual as a leader.

In this scenario, speed, and adaptability are crucial to success. Companies are urged not to become complacent once the first hints of normalcy appear. Instead, they must focus on recovery and resiliency strategies that are more sustainable. As the global fashion sector will be undergoing tremendous change, only then can businesses begin to understand what their "new normal" entails.

10.6 SECURITY AND CONSUMER PRIVACY WITH RESPECT TO ONLINE SHOPPING

With increasing internet use and online shopping, the risk of exposure to fraud has been rising substantially and efforts against this cyber-crime threat are deployed pan global level. Scammers impersonate real online sellers by creating a phony website or posting a bogus ad on a legitimate retailer's website. While many online retailers are reputable, scammers might take advantage of the internet's anonymity to defraud naive customers. Scammers utilize cutting-edge technology to create phony merchant websites that appear to be legitimate internet retailers. They may employ complex designs and layouts, maybe stolen trademarks, and even a domain name that sounds similar to that of a legitimate retailer:

1. **Warning Signs:** The following are some of the warning signals with respect to online consumer privacy:
    - Deep discounts or very low pricing with unbelievable benefits or quality assurance that sound too good to be true.
    - Request for immediate transfer of payment through electronic transfer. Some retailers may want you to pay for vouchers in advance before you can take advantage of a bargain or a giveaway.
    - The social media promotion platforms are very new and offer things at extremely low costs. Such retailers may have restrictive policies, and there are more chances of fraud at such websites.
    - An online retailer fails to give proper privacy, terms, and conditions of use, dispute resolution, or contact information. The vendor may be located abroad, or the seller may refuse to accept payment via a secure payment provider or a credit card transaction.
    - At times there are many hidden costs, such as the import duties in case who are ordering from overseas.

As one of the initiatives, the Government generally cautions the citizens against such fake shopping websites, asking them to be aware and conscious and thus help protect their hard-earned money.

2. **Consumer Privacy and Security Steps:** Some of the steps which can be adapted for safeguarding consumer privacy may include the following:
   - The customers must assure and check if the website or social media page has a refund or return policy mentioned in detail, and also that it is reasonable. In the event that something goes wrong, extensive complaint or dispute resolution mechanisms should be in place.
   - When shopping on the internet, customers need to know who you're dealing with. Investigate the merchant site and know if you're dealing with a domestic or international company, and if the things you're ordering will be shipped from another country, are there any additional tariffs, etc. 'Know your Seller' becomes important before you transfer money.
   - Only pay for things using a secure payment service (search for a URL beginning with "https" and a closed padlock symbol) or a payment provider such as PayPal or other recognized portals when making online payments.
   - Avoid any transaction with a stranger that requires payment in advance via money order, wire transfer, foreign funds transfer, pre-loaded card, or electronic currency such as Bitcoin. Money sent this way is extremely difficult to reclaim. Customers should never send their money to someone they don't know or trust, and never provide credit card or online account information to someone you don't know or trust.

Being a cautious and responsible customer is of paramount importance given the revolution that the retailing industry is seeing in times to come.

10.7 CONCLUSION AND POLICY IMPLICATION

According to a survey conducted by the National Retail Federation, 90% of the consumers have altered their traditional buying preferences. More than half of consumers have begun ordering things online mode. Furthermore, 60% of the consumers say they are scared to move out of their homes to shop due to the infectious virus. All these changes are making consumers prepared for a crisis situation resulting from the COVID-19 pandemic and are becoming more cognizant of health and safety regulations around them.

Consumers in lockdown situations have discovered more easy, economical, and accessible alternatives. Netflix, Prime, and Hotstar are examples of streaming services that are likely to divert customers away from movie theaters. This is analogous to ride-sharing services like Ola and Uber, which are more convenient than calling a taxi. Consumers may find it easier to work, learn, and shop at home as a result of the coronavirus. The new norms and requirements, such as wearing masks and maintaining social distance, would most likely change existing food shopping and delivery options. This is visible globally where people are now wearing masks and shields before going for shopping or taking public transportation. In the services industry, especially in personal services such as beauty parlors, physical therapies, and fitness centers, modified norms are now in place and are being followed. Music, culture, education, consultations, and businesses are now all flourishing online, with just a click away. The world has become closer than ever before.

Policymakers should create a more conducive environment for e-commerce enterprises to thrive. Businesses must act now to stabilize their operations, maintain continuity, and deal with the uncertain situation strategically. Customer and brand protection, short-term financial management, supply chain management, labor management and employee health, and legal and contract dispute resolution are among the services that businesses need to take care of. They should concentrate on the immediate future by strengthening enterprise resilience to ensure that they are prepared across all functions if the crisis persists. A government policy aiming at discouraging or encouraging spending is crucial in shaping future consumption patterns.

More screening and boarding processes, such as checking temperature, testing for the presence of the virus, and all safety measures will be part of the new normal. All major airlines are now implementing new procedures for passenger. Businesses are gradually adapting to the "new normal" that has emerged as a result of the crisis and are re-strategizing their business models to be in alignment with the new normal. Brands will need to align redefine promotions in accordance with their channels and put in place a revised product calendar to match fashion's "new normal" in order to improve their long-term sustainability. It is time for businesses to re-infuse value in products to appease their set of customers. The solution isn't just about decreasing overstock; it's about regaining the trust and enthusiasm of cash-strapped customers, which can't be done alone by discounting strategies. Businesses need to retain the customer by providing them a safe and pleasing environment to shop that is in accordance with the new normal.

Last but not the least; the internet is becoming a crucial power in the hands of the people. The way technology changes wants into needs has a big influence on the development of new habits like online shopping, online entertainment, and easy online education from the best institutions across the globe. There has been a switch from the traditional essentials which were food, shelter, and clothing, towards the new normal necessities which are education, transportation, and healthcare along with a phone, internet, and apps.

KEYWORDS

- **consumer shopping behavior**
- **COVID-19**
- **fast-moving consumer goods**
- **global economy**
- **impact on businesses**
- **over-the-top**
- **pandemic**
- **World Health Organization**

REFERENCES

Aytekin, E., (2020). *Steps Taken by Countries in Fighting COVID-19 Pandemic*. Retrieved from: https://www.aa.com.tr/en/health/steps-taken-by-countries-in-fighting-covid-19-pandemic/1812009 (accessed on 7 February 2023).

Chaudhary, H., (2020). *Analyzing the Paradigm Shift of Consumer Behavior Towards E-Commerce During Pandemic Lockdown*. Available at SSRN: Retrieved from: https://ssrn.com/abstract=3664668 or http://dx.doi.org/10.2139/ssrn.3664668 (accessed on 7 February 2023).

Coronavirus Impact on Cab Drives: Ola, Uber Cab Drivers Struggle, (2020). *The Economic Times*. Retrieved from: https://auto.economictimes.indiatimes.com/news/aftermarket/ola-uber-cab-drivers-struggle-amid-low-demand-extra-operating-cost/75852702 (accessed on 7 February 2023).

Covid and Consumer Behavior: What Will be the Impact of 2nd. (n.d.). *Zeebiz*. Retrieved from: https://www.zeebiz.com/small-business/news-Covid-and-consumer-behaviour-what-will-be-the-impact-of-2nd-wave-of-coronavirus-156765 (accessed on 7 February 2023).

COVID-19 has Changed Online Shopping Forever, Survey Shows, (2020). *Unctad*. from https://unctad.org/news/Covid-19-has-changed-online-shopping-forever-survey-shows (accessed on 7 February 2023).

Gaba, U., (2021). *Impact of the Coronavirus Pandemic on the Global Economy*. Retrieved from: https://countercurrents.org/2021/02/impact-of-the-coronavirus-pandemic-on-the-global-economy/ (accessed on 7 February 2023).

How COVID-19 Has Transformed Consumer Spending Habits, (2020). *J.P. Morgan*. Retrieved from: https://www.jpmorgan.com/solutions/cib/research/Covid-spending-habits (accessed on 7 February 2023).

How To Avoid a Scam | FTC Consumer Information, (2021). *Federal Trade Commission*. Retrieved from: https://www.consumer.ftc.gov/articles/how-avoid-scam (accessed on 7 February 2023).

Impact of COVID-19 on Consumer Behavior: Will the Old Habits Return. (n.d.). *Ncbi* from https://www.ncbi.nlm.nih.gov/pmc/articles/PMC7269931/ (accessed on 7 February 2023).

Kiran, R., K. M., & Nandha, K. K. G., (2020). Impact of COVID-19 pandemic on Zomato: A case study. *International Journal of Case Studies in Business, IT, and Education (IJCSBE), 5*(1), 14–24. doi: https://doi.org/10.5281/zenodo.4492571.

Lockdown Impact: Urban Company Acquires 33% New Customers, (2020). *Livemint*. Retrieved from: https://www.livemint.com/companies/news/lockdown-impact-urban-company-acquires-33-new-customers-11591859369313.html (accessed on 7 February 2023).

Lockdown Impact: Zomato to Lay off Employees, Opt for Temporary (2020). *Livemint*. Retrieved from: https://www.livemint.com/companies/news/lockdown-impact-zomato-to-lay-off-employees-opt-for-temporary-pay-cuts-11589535728636.html (accessed on 07 February 2023).

Lourenco, R., (n.d.). Ecommerce Fraud + 11 Fraud Prevention Strategies. *Bigcommerce*, Retrieved from: https://www.bigcommerce.com/blog/ecommerce-fraud/ (accessed on 7 February 2023).

Meyer, S., (n.d). COVID-19 Infographic: How Ecommerce Is Being Impacted. *Bigcommerce*. Retrieved from: https://www.bigcommerce.com/blog/Covid-19-infographic/ (accessed on 7 February 2023).

Singh, B., (2021). How the pandemic has Changed our Shopping Habits – Retail. *Livemint*, Retrieved from: https://www.livemint.com/industry/retail/how-the-pandemic-has-changed-our-shopping-habits-11620194521828.html (accessed on 7 February 2023).

The Evolving Consumer: How COVID-19 is Changing the Way We. (n.d.). *McKinsey Live*. Retrieved from: https://www.mckinsey.com/featured-insights/mckinsey-live/webinars/evolving-consumer-how-covid-19-has-changed-us-shopping-habits (accessed on 7 February 2023).

Index

D

E

N

O

P

Q

R

S